W9-BEX-930

CITIES
AND
NEW TECHNOLOGIES

ORGANISATION FOR ECONOMIC CO-OPERATION AND DEVELOPMENT

ORGANISATION FOR ECONOMIC CO-OPERATION AND DEVELOPMENT

Pursuant to Article 1 of the Convention signed in Paris on 14th December 1960, and which came into force on 30th September 1961, the Organisation for Economic Co-operation and Development (OECD) shall promote policies designed:

— to achieve the highest sustainable economic growth and employment and a rising standard of living in Member countries, while maintaining financial stability, and thus to contribute to the development of the world economy;

— to contribute to sound economic expansion in Member as well as non-member countries in the process of economic development; and

— to contribute to the expansion of world trade on a multilateral, non-discriminatory basis in accordance with international obligations.

The original Member countries of the OECD are Austria, Belgium, Canada, Denmark, France, Germany, Greece, Iceland, Ireland, Italy, Luxembourg, the Netherlands, Norway, Portugal, Spain, Sweden, Switzerland, Turkey, the United Kingdom and the United States. The following countries became Members subsequently through accession at the dates indicated hereafter: Japan (28th April 1964), Finland (28th January 1969), Australia (7th June 1971) and New Zealand (29th May 1973). The Commission of the European Communities takes part in the work of the OECD (Article 13 of the OECD Convention). Yugoslavia has a special status at OECD (agreement of 28th October 1961).

Publié en français sous le titre :
VILLES
ET TECHNOLOGIES NOUVELLES

FOREWORD

This publication contains the proceedings of a Conference held in Paris on the 26th and 27th November 1990 by the OECD Group on Urban Affairs in co-operation with the "Délégation française à la Ville" and "URBA 2000". It is derestricted on the responsibility of the Secretary-General.

ALSO AVAILABLE

Environmental Policies for Cities in the 1990s (1990)
(97 90 03 1) ISBN 92-64-13435-2   FF100   £12.00   US$21.00   DM39

Urban Housing Finance (1988)
(97 88 08 1) ISBN 92-64-13156-6   FF60   £7.50   US$13.50   DM26

Urban Infrastructure: Finance and Management (1991)
(97 91 07 1) ISBN 92-64-13584-7   FF95   £12.00   US$22.00   DM37

Prices charged at the OECD Bookshop.

*THE OECD CATALOGUE OF PUBLICATIONS and supplements will be sent free of charge
on request addressed either to OECD Publications Service,
or to the OECD Distributor in your country.*

TABLE OF CONTENTS

Introduction

Information Technologies: The Urban Dimension

Part 1

Information Technologies, the City, and the Urban Society

Part 2

Application of Information Technologies in Urban Services

Part 3

Making Good Use of New Technologies in Cities:
Facts, Reflections, Recommendations

A. **Testimonies by cities**

B. **Assessments**

STRUCTURE AND ARRANGEMENT OF THE BOOK

Introduction

Mr. Michel DELEBARRE, Minister of State, Minister for Cities and Land-Use Planning in the French Government, contributes the first of the introductory papers. He is followed by Mr. Jean-Claude PAYE, Secretary-General of the OECD. Mr. Emile BIASINI, Secretary of State for Major Public Projects in the French Government, is the final contributor to this section, discussing national and international policy dimensions.

Part 1

An underlying theme, which pervades many of the papers, is the potential for developing a new urban culture, associated with new information technologies and communication systems. This provides the basis for a grouping of four general papers which deal with the interaction and relationships of *information technologies, the city and the urban society* in various dimensions.

The first of these is by Mr. Gilbert PAQUETTE, Professor at Télé-Université, Montreal, and former Minister of Science and Technology in Quebec. He argues that we are approaching a second wave of the application of information technologies to cities. The first wave was concerned with gathering, organising and using information. The forthcoming challenge is to build a cognitive urban society, processing and applying knowledge gained from information.

The next paper in this group is by Mr. Gabriel DUPUY, Professor at the Institut d'Urbanisme de l'Université de Paris, and Professor at the Ecole nationale des Ponts et Chaussées. While reviewing present and future practice in managing the delivery of networks of public utilities – water supply, sewage and stormwater disposal, electricity supply, public lighting and security – he foresees the larger growth of integrated systems to control, deliver, and charge for public goods and services. He reminds us, however, that a minimum level of provision of such goods and services is now expected, and that important trends in the expectations and perceptions of people will help to mould this evolution.

A third general paper by Professor Lars QVORTRUP of the Telematics Project, Odense University, Denmark, reviews the progress of "telework", or the outposting of work by use of information processing and transmission. His own term for this is the development of work-related information transmission as a substitute for travel. He reviews the different definitions and the past and future definitions of telework and notes that there are still obstacles to the systematic recourse to this new type of work organisation.

The final general paper is by Mr. Bengt SÖDERSTRÖM, an architect and Director of a consultancy in Sweden, specialised in social development of local districts. He discusses information technology and localism: the interactions of systems supplying information for local needs. His review of case-studies shows how information systems are a vehicle of communication to bring together administrative imperatives with evolving and dynamic expressions of individual and community needs.

Part 2

Five papers deal more specifically with particular spheres or dimensions within which *information technologies have been applied to urban services.*

The first two deal with the development of infrastructure, in both physical and abstract terms, to aid economic development in cities.

Dr. Mark HEPWORTH, of the Centre for Urban and Regional Development Studies, University of Newcastle upon Tyne, United Kingdom, deals with the information services being generated by local authorities to address issues and information needs of enterprises arising from such trends as internationalisation, and the facilitation of transactions.

Professor Mitchell MOSS, Director of the Urban Research Center, New York University, provides a companion paper on how communication technologies have been deployed in hard infrastructure of different kinds to facilitate and accompany economic development.

Professor Neil WRIGLEY of the University of Wales, in Cardiff, discusses the use of information technologies in health care services, arguing that this is an essential accompaniment in changing existing public health financing and delivery systems into systems based more on market principles of "managed competition".

Dr. Jürg SPARMANN, of the Berlin branch of SNV, a study group for urban transport, research and development, presents a paper on new technologies to address travel demand management, with the particular objective of promoting the use of public transport. He reviews relevant experiences and realisations.

The final paper, by Professor Lorenzo CASSITTO of Politecnico di Milano University and Mr. Sergio SEGRE of SCS (Communications Strategies) of Milan, describes and assesses the use of information technologies and systems to monitor environmental conditions and improve energy use.

Part 3

This part, "Making good use of new technologies in cities", offers a series of testimonies, assessments and conclusions.

A. Testimonies by cities

This section presents the main ideas outlined by mayors and local officials from major cities and regions throughout the world. It provides background to the studies constituting Parts 1 and 2 by bringing a more political vision of the ways in which the new technolgoies are penetrating the day-to-day functioning of cities, and gives an idea of the expectations and queries that local officials have.

Mr. Michel GIRAUD, President of the Association of French Mayors and President of METROPOLIS, the World Association of Great Metropolises, pleads for the use of telecommunications in order to expand exchanges of all kinds and indicates how "Greater Paris" is moving in that direction.

Mr. Antonis TRITSIS, Mayor of Athens, and Mr. Jordi BORJA, Deputy Mayor of Barcelona, believe that the new technology should help to extend local democracy and social cohesion. They both express the wish that "networks of cities" will be established so as to exchange information and experience.

Mr. Arthur EGGLETON, Mayor of Toronto, outlines the efforts his city has made in recent years with regard to health and the environment so as to make and keep Toronto a "healthy" city.

Mr. Jaap ENGEL, Director of the Foundation for Information Technology, in Amsterdam, explains how the city intends to take advantage of modern data-processing and telecommunication systems in the context of an overall competitive strategy intended to enhance the city's attractiveness as a business centre.

Mr. Toshiaki NINOMIYA, Deputy Mayor of Osaka; Mr. Yasuhei SATO, Deputy Mayor of Yokohama City; and Mr. Shinichiro SHIMAMURA, Mayor of Koshigaya, outline their plans for "intelligent cities" in Japan based on intensive use of telecommunications.

Mr. Werner HAUSER, Chief Executive Officer of the Association of Local Authorities of the State of Baden-Württemberg, Germany, emphasizes the progress made in streamlining the functioning of German cities since the penetration of information technology and telecommunications.

In a more futuristic vision, Mr. Gianfranco DIOGUARDI, President of Bari Technopolis, Italy, describes an "enterprise-city" project involving "neighbourhood laboratories", the result of co-operation between the Technopolis, a construction firm and a training organism; and Mr. Paul BARRATT, Special Adviser to the Australian Department of Foreign Affairs and Trade, outlines the "Multifunction Polis" project to be carried out, in co-operation with Japan, on a site at Adelaide.

B. Assessments

Professor Peter HALL of the Institute of Urban and Regional Development at the University of California-Berkeley provides a paper on the theme of new technologies, participation, integration and lifestyle: a subject reflecting the underlying awareness that we stand on the threshold of a new information society and urban culture.

Dr. Costis TOREGAS, President of Public Technology Inc. (PTI) of Washington, D.C., summarises discussion concerning new technologies and public sector management, concluding that the crucial question is who controls and directs the processes involved. He coins the term "orgware" to cover software and hardware plus organisational questions.

The final review of the use of new technologies and local economic development is by Professor Rémy PRUD'HOMME of the Institut d'Urbanisme de l'Université de Paris-Val de Marne. He maintains that a fundamental issue is whether new technologies actually tend to reconcentrate or decentralise economic development.

C. Conclusions

Mr. Georges MERCADAL, President of URBA 2000, singles out the key ideas emerging from the studies, testimonies and assessments outlined during the Paris Conference and formulates recommendations and proposals (these were adopted by the participants). Dr. Siegfried BRENKE, Head of the Urban Affairs Division, OECD, provides the concluding remarks.

Introduction

INFORMATION TECHNOLOGIES:
THE URBAN DIMENSION

INFORMATION TECHNOLOGIES:
AN OPPORTUNITY FOR THE CITIES

Michel DELEBARRE
Minister of State, Minister for Cities and Land-Use Planning
FRANCE

The "New technologies", a blend of computer science and telecommunications, are accelerating the transfer of information to a fantastic extent. They have practically abolished space and time.

What does this hold in store for our cities? Throughout the world they are already feeling the impact of the transformations brought about in the modes of production, trade and distribution.

First of all, as producers of services, cities will find that the use of the new information technologies can make them more efficient at equal or lower cost. The many examples described in this book prove that in every area of urban responsibility, cities can respond to their inhabitants' ambitions more fully and without greater expense – thanks to these technologies.

But that is not all. Two fundamental questions arise. The first has to do with the ways in which the development of these new technologies will affect the localisation of activities: will telework drain our cities of at least part of the population who until now have lived there, handling and disseminating information? Or will the big cities exert, on the contrary, a more magnetic influence than ever before? Developments in recent years would seem to confirm the second hypothesis even though the new information technologies do offer opportunities to everyone, not only in the great metropolises but even in rural areas as well.

The second question has to do with keeping the conditions in which these new technologies develop under control. Public bodies, and local authorities in particular, are too directly and too deeply concerned to allow these technologies to proliferate haphazardly, guided only by considerations of supply and availability. Whatever the technical aspects may be, it is democratic considerations that must have the final say. Moreover, the new information technologies can be used to help combat the process of marginalisation in our cities and outlying districts. By providing specific services right in the home they can help to reinsert the most destitute or underprivileged into society. Convincing experiences are there to prove it, and ample testimony is found in this book. We must seize the opportunity, broaden the scope of our experiments and research, and strive to advance from the particular to the general.

The initiative taken by the OECD and its French partners – la Délégation interministérielle à la Ville and URBA 2000 – deserves our full attention for it shows that regardless of the countries concerned, the challenges that cities are currently facing with regard to the use of new technologies are the same everywhere.

The dialogue which has been launched among cities will enable them to strengthen their co-operation, consolidate their efforts, find solutions and develop new ways to share and disseminate their ideas. I have the fullest confidence that this will be the case.

TECHNOLOGICAL CHANGE AND URBAN DEVELOPMENT: AN INTERNATIONAL CHALLENGE

Jean-Claude PAYE
Secretary-General of the OECD

The sweeping economic and social changes that are taking place today offer cities new opportunities for development. They also raise problems. While the cities in which three-quarters of the OECD countries' population live reflect these socio-economic and demographic upheavals, they are at the same time the crucible for change. They remain the principal symbol of civilisation. It is in the cities that growth and innovation are generated. But it is there, too, that the social problems that affect, for instance, young people, old people and the ethnic minorities crystallise. And cities are the main sources of environmental problems.

For a very long time urban policies were characterised by a compartmentalisation that reflected the division of responsibilities among ministries and among national, regional and local authorities. And yet it is in the cities that economic, social and ecological systems are most interdependent. It is in the cities that it is especially necessary to see to it that the activities of the different sectors are in synergy rather than in conflict with each other. Many of the OECD countries are now trying to apply more closely co-ordinated policies to shape the development of their cities.

An intersectoral approach to urban policy implies the ability to bring together all the actors concerned so that they can contribute jointly to defining integrated action for development.

The seminar is timely, too, in that today there is greater awareness of the scale and complexity of the relations between technological change and urban development. Modern information and communication technologies, in particular, are changing the world we live in, and the cities most of all.

Technological progress over the past two decades has not always had the happiest consequences for the economic development of our cities. Technical innovations and better communications have enabled industrial enterprises to set up far from cities, and this has drained from the urban areas the economic vitality that stemmed from the availability in one and the same place of skilled labour, services, transport facilities and markets. Thus change has sometimes caused cities to lose out, and revitalising them is a key priority. Carefully considered use of the new technologies can make a substantial contribution to that revitalisation.

The OECD has undertaken a very broad-based study on the linkages between technology and the economy. That study is now entering its final phase. It highlights the essential role of the "site" in the process of developing an disseminating technological innovation. Innovation canno flourish in a desert, even one well supplied with oases. It calls for a concentration of very different actors, able to benefit from the economies of scale and the synergies on offer on the right site, that is to say one that is welcoming and well-equipped. The role of the site is all the more important in that the human resources that are the key factor in innovation are increasingly mobile and tend to congregate in the most attractive places. Those places in turn reap the fruits of innovation, and this interaction consolidates human and economic bases alike.

It seems that this point of ongoing interaction is one that should be stressed when we consider the impact of the new technologies on the economic development of cities.

If urban infrastructure is to be of high quality, more and more of the equipment needed for information and communication purposes has to be made available. That equipment can make public services much more efficient. In particular, as the work of the OECD's Group on Urban Affairs has shown, the application of the most up-to-date networking techniques and the integration of software are opening up fascinating prospects in public works planning, resource management, the definition and delivery of urban services, the simplification of procedures and a host of other areas. All of this may substantially boost the productive capacity of urban economies.

On the social front too, information and communication technologies open up new opportunities. Social justice demands equality of access to information and the means of education. Thanks to the new technologies and the possibilities they offer – such as individually tailored training packages and videotex services – the technical barriers to the wide diffusion of information are done away with and so, to a large extent, are the economic barriers too. This means that one of the main curbs on social justice is tending to disappear, and that the ordinary citizens will henceforth be able to play a greater part in local life. The very fact that information on social, cultural and scientific matters is becoming more plentiful and more widely available enhances the quality of life for each individual.

Lastly, a few words about one question that is now very much in the minds of the OECD's Member countries – the environment. The mandate of the Group on Urban Affairs requires it to look at ways of reconciling the imperatives of economic growth with those of environmental protection. Furthermore, the OECD has just launched a new programme on technology and the environment. Of the many projects it comprises, two – one on urban transport and one on best practices for urban energy management – have a direct bearing on your work here today.

By definition urban policies have to do with the long term. Strategies that are to run for 10, 20 or 30 years have to be grounded on assumptions that are necessarily based on state-of-the-art knowledge. Underestimating the scale and importance of technological progress is a mistake that has often been made in long-term forecasting. Harnessing the new technologies in the service of urban development is thus a tricky task for those who have to frame urban policies at local, regional or national level. Discussion at international level may be helpful. This Seminar will make a substantial contribution to that end.

TECHNOLOGICAL CHANGE
AND EQUAL CHANCES FOR URBAN DWELLERS

Emile BIASINI
Secretary of State for Major Public Projects
FRANCE

"Cities and New Technologies" are a matter of personal involvement both for the President of the Republic and for the Prime Minister, since, after all, re-inventing the city is among the most pressing of the challenges facing society today.

The only way to cope with the relentless demographic thrust is to keep on toiling day by day, in pursuit of an ideal, that of achieving a balance, or at least of matching our vast strides in technology with their consequences for human life. One of the aims of this conference is indeed to identify "the qualitative and quantitative benefits that new technologies offer local authorities".

Quantitative benefits should be measurable enough, but what about the qualitative ones? Perhaps we could say that those add up to the well-being of our city dwellers.

It is certainly reasonable to ask how far peoples' well-being has been really commensurate with the spectacular advances in technology. No need to stress their material consequences, affecting every aspect of community life, from communications, traffic safety and transport to services, health and education.

Perhaps, though, these advances are opening up new divides between those who can reap all their advantages and those who cannot, between those who can understand them and those who cannot, between the powerful and the powerless.

But we also have to adapt to change.

Before the war Paul Valéry, awarding prizes to the pupils at a school in Sète, used an amusing fable to illustrate the need to adapt. What, he mused, would such great geniuses of the past as Archimedes, Newton, Galileo or Descartes have been able to make of a dynamo? Their bafflement would raise a chuckle or two. How much more amazed they would be today if they were shown a microchip!

Valéry concluded by telling the pupils that one of the inescapable and most cruel effects of progress was the way it added a penalty to death itself, a penalty that could only grow in severity as the revolution in customs and ideas gathered pace.

Today that extra penalty surely lies in the dichotomy between the annihilation of time and distance barriers and the development of each individual for whom that same time remains biologically immutable. Gestation, everything we gradually learn and experience, putting together that "pitiful hoard of secrets" which make up a person, still require an irreducible period of time.

As advancing technology defies time, opening up direct communication between the iron age deep in Africa, the middle ages in some nearer countries and the provocative era of western consumerism, it is generating new forms of stress. Where does today's man stand, let alone the men of any other era?

How is he to find the time it takes for the many streams fertilising his plain to deposit their sediment? Isn't that, when we really think about it, where our young people's crisis is rooted? How can technology offset the imbalances it is itself creating? The cities are the obvious place where technology growth is exerting its most direct influence. There is no need to spell out all the benefits.

But in overcrowded cities people are constantly brought face to face, which produces as much exasperation as emulation. It is not true that the city causes the crises, their causes are at a far greater remove, but the city does bring together the factors. This is where a sick society sees its sicknesses break out. Which is why the problem of the cities has become the problem of civilisation itself.

The technological benefits more readily available in a city make it more of a magnet. But the technological developments essential to its growth and survival also aggravate the risks and the dangers. Surely what our new towns and districts actually represent is a challenge to time, to the gradual maturing of a social balance which can only achieve its patina with the lapse of time. Today, though, we can no longer wait for that to happen and therefore have to devise some kind of alternative.

Technological evolution may be welcome, providential, inevitable and desirable but it also needs to generate the remedy for its own dangers. Technology enables man to communicate more easily, helps to improve living standards, promotes fuller relationships, material and social advance. But communication assumes some message, and the message needs to advance just as inevitably as its medium. Not much will be achieved by amplifying the signals if most people are still going to find them incomprehensible.

Our cities are certainly not responsible for the disorders they are merely bringing to light. But their leaders do have an obligation to offset sources of inequality and so obviate the resulting disruptions. We sometimes have to look on, astonished and powerless, as violence flares up in some area where every effort had apparently been made to relieve tensions at the root of a malaise which resists any kind of logical approach, so much harder is it to formulate than technological advance. Technology is rational, the malaise is utterly irrational, and so our cities face the difficult task of achieving progress simultaneously in both those disparate but wholly interdependent universes.

City councillors understand the problem better than anybody else, and know that it can have only one solution, a tenacious, unremitting attempt to improve equality of opportunities. That is the basis of democracy, and it is something best learned in a city. But they will also be aware that with a problem like this, nothing seems to have been achieved until everything has been achieved, and that unfortunately, one can never have achieved everything. That is why the attempt really does have to be unremitting. The resources it demands are of course both technological and financial. But what it mainly demands is something that neither technology nor finance can contribute, it demands generosity.

For at the same time as we are developing the technologies with which to bring people closer together, a gap seems to be opening up which technology can do nothing about. It is certainly within each individual that the ground needs breaking, by means of what we call cultural activity, and the technologies ought to contribute here, especially the vast facilities for intercommunication, and television in particular.

Ultimately the necessary, dizzying development we want for technologies at the service of our city communities only makes it all the more essential to press ahead with that other, irrational form of progress which must perhaps just be called happiness, one which people are increasingly clamoring to be able to attain.

Probably not by chance, these ideas about future progress lead me naturally back to the past and to something that will last, perhaps because it is timeless, to a point made by Rabelais. "Science without conscience", he said, "can only ruin the soul."

Part 1

INFORMATION TECHNOLOGIES, THE CITY, AND THE URBAN SOCIETY

THE KNOWLEDGE CITY OF THE YEAR 2000: CONSTRUCTION OF EDUCATIONAL AND CULTURAL NETWORKS

Gilbert PAQUETTE
Professor at Télé-Université of Quebec
CANADA

Gilbert Paquette has a masters degree in mathematics, a masters degree in informatics and doctorate level training in artificial intelligence from the Université de Montréal. He is currently preparing a doctoral thesis on the educational applications of artificial intelligence at the Université du Maine (France).

He was president of the Quebec Association of Mathematics, and Quebec's representative in the international commission at the origin of the modern mathematics movement. Today he is professor of mathematical didactics and informatics at the Télé-Université of Quebec.

Elected to the National Assembly of Quebec in 1976, he became Minister of Science and Technology (1976-1985).

Since 1985, he has concentrated on researching the educational applications of artificial intelligence. In this context, Gilbert Paquette has helped design some ten software packages and participated in their experimental use in schools, as well as helping develop a new system for designing software teaching aids and various educational applications of learning-based systems.

> *"We are entering a new universe where the real raw material is grey matter; and by far the most important production is that of knowledge and information"[1].*

"End twentieth century demographic, economic, social and cultural realities demand a new and different educational system"[2].

FOREWORD

In this paper on the educational and cultural dimension of the cities of the year 2000, if one word, "knowledge", crops up all the time it is because we firmly believe that the second wave of information and communication technology (ICT), involving the processing of knowledge, is going to bring in its wake tremendous changes in the way we produce, consume and decide what direction to follow – as well, of course, as in the way we acquire, process and diffuse knowledge. The first wave carried us into an information society. The second will carry us into a cognitive society, cognitive cities, whose fabric and sub-systems will rely as never before on knowledge.

Introduction: needs and resources of the information society

The new technologies of the information society create new educational and cultural needs, just as they place new means at our disposal and offer new solutions. The city, the social system nearest to the citizen, is at the heart of all these changes because it is mainly there that the educational and cultural patterns of tomorrow's society will be formed and shaped – via education systems, on-job training, consumer information processing, and services and facilities catering for cultural and leisure pursuits. We shall give an account of these patterns and in doing so highlight some of the ways in which it seems to us that change should be channelled, as well as giving some examples of trail-blazing achievements.

From the outset our implicit assumption is that the city will continue to be the core of community life. Even if its geographical spread becomes more extensive, as is sometimes forecast, the need for urban focal points where knowledge-related economic, social and cultural activities can be housed, organised and encouraged will be all the more pressing as the city's cognitive (and affective) fabric becomes stronger and closer-knit – as we believe it will.

For most of the four million years of his existence, the only means of communication available to man was the spoken word. Until the Sumerians invented the first known system of writing little more than four thousand years ago, information and knowledge could be passed on only by word of mouth by one person to a few others within earshot.

During the greater part of history since then, writing as a means of setting down and communicating knowledge was inaccessible to ordinary people. Until the printing press was invented, a little more than five hundred years ago, only a privileged elite, a handful of scholars, was able to draw on the remembered knowledge stored in great libraries and so contribute, through the written word, to increasing the sum of that knowledge.

And even during these last five hundred years, although knowledge has been accumulated and passed on to more and more men and women, the means used have been extremely limited, compared with those that are available today in the information society.

The information society, a reality

The task of acquiring, processing and passing on knowledge remained mostly a matter for a minority of scholars, teachers and writers until about 40 years ago. Then, in the early 1950s, thanks to the impact of television and the new information and telecommunications technologies we moved on very rapidly from the industrial society to the information society. Today, in the post-industrial societies of Europe, North America and Japan, more than 60 per cent of the labour force is mainly concerned with the processing of information and of knowledge.

Mounting pressure on education and cultural services

In the post-industrial society information and knowledge are not only the principal raw material but usually the principal product too. In particular, the successful operation of businesses, cities, government and so on increasingly depends on ability to pick out from the ever-increasing mass of available data only the information that is really of value, and to channel it into productive uses.

Other broad trends typical of the post-industrial society also converge in this general direction and shape the pattern of educational and cultural requirements; in particular the constant changes in job content and in the working environment call for new forms of initial and continuing education and training. At the same time, ever-growing demand due to shorter working hours and the increasing numbers of active retired people put pressure on educational and cultural services. What is more, the increasing numbers of people working at home and producing for their own consumption want distance information and learning to be available in their homes so as to be able to receive the "raw material" they need, process it on the spot and transmit the results elsewhere.

Turning information into knowledge

Happily we can look forward to meeting these new educational and cultural needs, thanks to the fast and integrated development of ICT, hyper-media and telecommunications networks. A first wave of ICT, from 1950 to the present day, took over some simple and repetitive tasks and dealt with a very limited amount of knowledge. Now, with the second wave, knowledge in all its forms can be computerised and transmitted all over the world.

The biggest change, and the one that holds out most promise for the future, is that computers today can handle and process information in such a way as to infer new knowledge from what is already known. Work on artificial intelligence now has practical applications. Natural language can be processed, forms recognised, *knowledge-based expert systems set up.* The new information and communication technologies also make it possible to perfect the most commonly used tools – for instance, spreadsheets, databases, text and graphics processing programmes – so that more and more "intelligent" tasks can be performed and higher-level knowledge handled. Such new software makes the computer even more valuable as an instrument for educational and cultural purposes.

Even more possibilities are being opened up with the arrival during the past two years of *hyper-media* workstations using a combination of ICT so that knowledge can be accessed and processed faster. With multi-media software, digitised data, text or graphics can be handled as easily as video or audio tapes. Information is stored on videodisk or CD-ROM to constitute a database on a given subject, with an interactive programme for non-linear access.

Gradually, too, optical fibre *telecommunications networks* that can transmit sound, images and computerised data to anywhere in the world are coming into general use, all this information being managed by the new, more "intelligent" computers. The development of telecommunications services is moreover largely dependent on modern information processing technology without which it would take too long and be too complicated and costly for the ordinary person to obtain the information he wants.

In this second wave of ICT there is more and more talk about "networking knowledge". Knowledge stored in different media is digitised and either processed on the spot or teleprocessed, using "intelligent" methods akin to human thought processes. Clearly this technological revolution is already having and will continue to have far-reaching consequences on the educational and cultural front.

State of the art: ICT applied to education and culture in the cognitive city

This rapid overview of the main educational and cultural needs of the information society leads us to look rather more closely at some of the information and communication technologies that can already be applied in preparing the way for education and cultural life in the cities of the year 2000. We first identify five main fields: school, training at the workplace, learning at home, media resource centres and museums, and alternative community services, and then go on to try to form an overall picture of the acquisition and processing of knowledge, encompassing all these fields.

Processing knowledge at school

The report on an OECD CERI Conference held in 1986 highlights a major change of direction in teaching children the basic skills traditionally summarised by the terms reading, writing, mathematics and science. The conference noted that:

"Our children must learn to acquire information, to communicate, to reason qualitatively and quantitatively and to solve everyday problems. They especially need the self-awareness skills that will permit them to continue learning on their own... This is the dilemma. Providing high quality basic skills education sufficient to meet the needs of a long life in a rapidly changing world is expensive and may require skills that are not always present in those willing to be teachers... Two forces have emerged that might enable us to solve this dilemma: inexpensive information processing power and a blossoming cognitive science that can rigorously study, understand and improve the educational process"[3].

The OECD specialists thus upheld the cognitive theory of information built up as a result of studies on artificial intelligence carried out over the past 30 years. This theory is now firmly enough established to serve as an approach to the educational and cultural needs of today – and of tomorrow. It also provides a theoretical basis and practical problem-solving tools to ease the transition to a more broadly "cognitive" information society.

Bringing education systems into the post-industrial age

Currently, education in all countries is largely of the industrial type. It has to be acknowledged that educational tasks are "Taylorised". Broadly speaking, curricula, teaching methods, agreements on working conditions and rules governing budgets are based on a sort of received truth. Education still consists of collective instruction rather than individualised teaching. It is essentially the same for all, provided by teachers specialising in a given subject and summarily assessed in the same manner and at the same moment.

For example, in a survey published in 1989, the Conseil Supérieur de l'Education du Québec reported that:

"Rather more than 80 per cent of primary and secondary schoolteachers said they practised group teaching in which the content was the same for all the pupils in a class, with hardly any allowance being made for different learning rates. On the whole, teachers 'chalk and talk' with very little reference to everyday life and tend to teach subjects in isolation one from another"[4].

Patently this type of traditional instruction is ill-suited to fostering the new basic skills needed today, which call for individual attention and more personalised teaching with a view to helping the pupil to build on his knowledge for himself. The child, guided by the teacher, searches out information, processes it in the light of the tasks and projects to be carried out, and then uses it to solve real problems or to procure further information. So teaching on traditional lines is particularly unsuitable to the needs of the post-industrial society, which requires people to be able to acquire information and knowledge for themselves and "learn to learn".

Experience in many countries over the past ten years has demonstrated the potential – underused though it still is – generated by the first wave of ICT. Microcomputers and distance learning are transforming education systems and radically changing the learning and teaching process in line with the requirements of the information society.

Rational use of the tools offered by the second wave of ICT with its "intelligent", interactive hyper-media will continue and intensify this change, since "intelligent" computers can now use different media to represent, process and send wherever required not only raw data but also higher forms of knowledge.

The approach proposed here is a pragmatic one, which assumes that it is possible, thanks to the new ICT, to adapt teaching methods to the post-industrial era without changing the school and its environment as currently organised, *i.e.* as a collective system functioning in a classroom with a teacher and about 30 pupils, clear-cut targeted curricula, a central authority responsible for educational content and administration and decentralised management at school level. We may wager that in the long run this type of organisation will appear increasingly anachronistic and that it will be modified to the extent that teaching is radically and successfully reformed.

The aim: Training in basic skills

Education systems must now focus on learning the new basic skills identified in the OECD report quoted above:

- acquiring knowledge: searching for information, finding one's way around a specific area of knowledge, pinpointing and extracting, decoding, representing, structuring, fitting together;
- processing knowledge: through analysis and synthesis, through induction and deduction, through calculation, graphic representation *etc.*, assessing what one has learnt and adding to it;
- applying knowledge: working out possible solutions, looking for the rules that apply, taking decisions, planning, using problem-solving heuristics, assessing results;

– passing on knowledge: choosing the right piece of information, summarising it, setting it out in a way that will suit those for whom it is intended, putting it over by means of the appropriate media.

We have to start by overhauling primary school curricula, because if children are to learn how to learn for themselves they have to start young. Obviously educational content will remain largely unchanged (mother tongue, mathematics, the natural and human sciences), but subjects will be taught in an integrated and interdisciplinary manner. Subsequently, at the different school levels, training in basic skills will be pursued in *knowledge-processing workshops*[5] and supplemented by more specialised subject teaching as pupils progress.

Helping the teacher to help the learner

As Denis P. Doyle points out, *"in the industrial school model, the teacher is the worker and the pupil is the product... In the post-industrial school, the teacher is the manager of the learning process and the pupil is the worker"*[6].

But both teacher and pupil have to be given the tools for the job.

Sooner or later we shall no longer be able to do without ICT as a teaching resource – especially as software, increasingly shaped by what we now know about human cognitive functioning, is becoming more powerful and at the same time easier to use. The new information and communication technologies, interactive and usable for managing other media, provide a tremendously valuable tool for teaching and learning. Everything that has to be learned can be fitted into one or more software packages. Pupils can work on the system on their own or in small groups. And the computer frees the teacher to concentrate on his main role, which is to help the pupil to learn.

There is no country in which the vast apparatus of state education is not in need of new impetus and long-term planning to adapt to the needs of the post-industrial society – giving curricula a knowledge-processing bias, building up multi-media databases in the various fields of learning, establishing activity banks and devising teaching tools, modernising initial education and upgrading teachers' skills on an ongoing basis, providing teachers with counselling support as the new technologies are brought in and, of course, gradually giving schools the hardware they need.

Further training at the workplace

Training at the workplace, too, has to meet the needs of the post-industrial age. This is a prerequisite if businesses are to remain competitive and organisations efficient.

The quantity and quality of the knowledge to be acquired

In the information society, the kinds of knowledge businesses and other organisations require in order to function efficiently change very rapidly. This means that the nature of tasks and interpersonal relationships within an organisation is also constantly changing, creating the need for continuing training at the workplace.

The training that is required changes too. As well as merely acquiring factual knowledge and learning how things function, those working in a technological and organisational environment that is constantly changing have to take on board new transferable concepts, processes and methods.

"Taylorised" vocational training

Every year governments and businesses in the post-industrial societies spend fortunes on training; but most of the money goes on taking staff away from their usual jobs in order to train them, bundling them off somewhere to a class that closely resembles the industrial school model, the main difference being that training is short and specifically targeted on, say, learning a new process, using new equipment, or becoming familiar with a new method, once and for all, if possible.

It will not be possible to continue this type of training for much longer – training with no follow-up, snatched at odd moments and divorced from the work context. It ought to be possible, by bringing into play the interactive tools made available by ICT, to provide hands-on training at the workplace while the work in question is actually being done, thus motivating the students by bringing them face to face with the problems that really do have to be solved on the job and making the learning process part of the job.

Gearing training to the task in hand

Furthermore, the time is not far off when it will be necessary to gear training to the task in hand so that the former changes with the latter. To achieve this end, it must be possible to change training tools quickly and easily, preferably in agreement with their users. This means the tools have to be flexible and user-friendly. As emphasised by the instigator of one of the projects we studied:

"The rapid turnover of skilled workers entails recurrent expenditure of energy on training and supervising them... Unlike written instructions, which one cannot always be consulting in the everyday hurly-burly of the workplace, the conversational mode and the ease with which knowledge can be adapted via an expert system make easy reference possible and provide an apt answer in every situation. Experimental installation of knowledge-based expert systems has shown them to be promising tools for training" [7].

Resources for the development of vocational training

Imagine a company excusing itself for not being able to keep up with events by saying lamely that the world has changed. It would be laughable. And yet training is still probably the function most neglected by business and administration.

The sums spent on it by North American and European companies are ridiculously small compared with their total sales figures, particularly when it is realised how vitally important human resources and the skills at their disposal are from an economic standpoint.

Governments, in consultation with industry, should seek to provide the right climate for a thorough overhaul of training in the workplace. Awareness is dawning that such an overhaul is long overdue. Vocational training needs practical support, for example joint funding by central and local government and business. And firms and state-run vocational training institutes should jointly set in train innovative model training schemes for use in different economic sectors.

Learning at home

Nowadays it is quite common for people to produce goods for their own consumption or to work at home, and this, coupled with the fact that everyone is living longer and has more spare time, has led to a demand for educational and cultural services at home. This demand is slowly but surely being met, in a variety of ways.

Educational and cultural television

Television and cable networks increasingly feature one or more channels entirely or partly given over to educational and cultural broadcasting or videotex. In most countries you can now improve your mind without enrolling in any institution, simply by sitting at home and watching television.

Formerly a passive medium, television is becoming interactive. By using a cable-selector it is now possible to choose what is shown on screen. One interesting application puts together news bulletins which are then viewed as summaries of fuller reports. By zapping with a remote control, the user can call up the full report on a subject that especially interests him. Documentaries and other educational programmes could be put together in the same way.

Videotex services for the general public

Numerous videotex services distributed via cable or telephone links now allow home microcomputers to be operated as online terminals for accessing databanks. Software in the public domain, including some excellent tutoring systems, can also be transmitted via videotex to microcomputers in users' homes.

One particularly useful training application for videotex is teleconferencing, which allows a group of people, who may even be in different countries, to communicate with each other. Each member of the group uses his microcomputer to receive data from other members and to transmit replies at his convenience. Users can thus establish discussion groups that are free from the constraints of time and distance.

Distance education

In a number of countries students can earn credits towards university diplomas by enrolling on distance learning courses run by ''open'' universities which use multi-media, learner-based instructional material that is either sent through the post or transmitted directly to students' homes. The material is presented in the form of an assisted course of self-tuition that enables students to gain access to knowledge stored in a variety of media without the help of a conventional tutor but with back-up from course supervisors.

A significant change has taken place in distance learning since the 1970s, when there was a shift in emphasis away from correspondence courses towards the use of multi-media. A study investigating this change noted in its conclusion that: *''Demand (for distance learning) is growing rapidly in both Europe and North America: people are looking for new systems of continuing education that offer flexible course structures, alternative forms of instruction and value for money. The new communication technologies are expected to play a major role in meeting this demand, and in view of the commercial success that they have already achieved the future for distance learning would appear to be bright on both sides of the Atlantic. This worldwide growth in distance education has been accompanied, at least in some parts of Europe and North America, by fairly significant changes in the teaching models adopted within education in general. These models encourage the use of teaching strategies which are less prescriptive, less concerned with the formal transmission of knowledge, and which therefore allow greater scope within the educational process for independent work by the learning adult''[8].*

Distance education is clearly making increasing use of the tools and methods afforded by the information society. Despite this trend, however, videotex services, interactive television, knowledge-based software and hyper-media have yet to be fully integrated into distance learning systems. Indeed, there are some studies which would seem to indicate that growth in distance learning will be fuelled not so much by technological improvements as by the political and socio-cultural factors that make education and the acquisition of knowledge a high priority in people's minds.

Media resource centres and museums as the focal point of educational and cultural networks

The increased availability of educational and cultural services within the home through television, cable networks and videotex services does not dispense with the need for major cultural facilities for the public – quite the opposite in fact. In the same way that the advent of television has had a positive impact on the theatre, the transmission of knowledge into the home is likely to have a similar knock-on effect in terms of the use made of museums, interpretation centres, libraries and media resource centres. One of the main results of the information society is that people will have ever-increasing amounts of free time to devote to leisure activities.

The media resource centre of the year 2000

In the meantime, cultural facilities have already started what should prove to be an ongoing process of change aimed at capitalising on the potential benefits of the new information technologies. The change from the processing of information to the processing of knowledge is of critical importance to the design of the media resource centre of the year 2000.

Over the last 20 years libraries have computerised their documentary references, which have then been integrated into networks to allow users to carry out online bibliographic searches. In most countries it is possible to use the terminals now installed in a large number of libraries, usually university or public libraries, to search for titles by subject or author.

Furthermore, libraries are now tending to store a broader range of media in their collections. In addition to books, libraries have collections of films, video cassettes, slides and even software. The term "media resource centre" is increasingly being used in place of "library".

These two separate development paths are steadily converging towards the remote distribution of not only references but also actual documents – and not only textual matter, but also still and animated images and software. This trend will start to gather pace once the cost of using hyper-media technology falls to more reasonable levels. Once that happens, the whole question of interaction between media resource centres, personal computers, distance learning networks and teaching systems, both in the workplace and in schools, will assume a new dimension.

The museum facilities of tomorrow

Modern cultural facilities – media resource centres, museums and interpretation centres – should gradually be integrated into a shared pool of knowledge that would be made available not only to visitors, but also, through videotex links, to users in schools, the workplace, and the home.

What distinguishes a museum from other cultural facilities is the presence of authentic or fabricated objects illustrating the theme of the museum. A traditional museum will concentrate solely on a single collection of exhibits: archaeological finds, examples of technology, works of art *etc.* The visitor is a passive participant who simply looks at exhibits which are described by a brief label. Sometimes the museum shop sells a guidebook that gives the visitor a detailed account of some of the items in the collection.

The current generation of museums offers a new approach based on discovery and interactive dialogue initiated by the visitor. In discussing the new approach taken by science and technology museums, Victor J. Danilov stresses that: *"All have a basic philosophy which distinguishes them from traditional museums and turns them into effective vehicles for informal scientific training. The crux of this philosophy is that the acquisition of knowledge about science, technology and industry can be enjoyable, and that the most effective way of learning is to use contemporary interactive exhibits"[9].*

A wider role is assigned to different media: *"A multi-media centre. Given that an interpretation centre, unlike a museum, does not make use of self-explanatory specimens/objects, it can accommodate full-scale media integration. The cultural message within the centre will therefore be communicated by a variety of means that will include descriptions and historical re-enactments: reproductions, interpreted reproductions, animated models, spoken replies, slides, videos etc. We feel that such variety is essential if the centre is to sustain the interest of visitors and in particular if it is to achieve the intensity and effectiveness required... Written explanations... will be kept to a strict minimum in order to give the centre a visual rather than a textual feeling"[10].*

The majority of recent exhibitions have managed to integrate the setting for exhibits into an audio-visual environment. Computer-based teaching materials are used freely and allow visitors to gain a deeper insight into certain aspects of the subject through relaxed, interactive dialogue. They have yet to become an integral part of the main exhibition, however.

The second wave of information technology will help to pave the way for interactive media integration in both time and space. IT methodology, in terms of data acquisition and processing resources, will allow organisers to create an intelligent museum. The museum will be able to convey knowledge in a manner that is user-friendly, varied and versatile, using new interfaces that will free users from dependence on keyboards and even mice, while providing access, through hyper-media, to a world of sounds, images and objects by means of non-linear excursions into the world of knowledge.

In conclusion, we can expect to see ever greater use of information technology in the museums of tomorrow. Applications will be many and varied:

– acquisition and updating of knowledge – the collective memory of the museum;
– non-interactive management of complex sets of media, as in the case of large-scale multi-media presentations;
– support for interactive, motivational games; it will be possible, for example, to control interaction whereby participants initiate changes in their environment by actuating lighting and audio-visual projections;
– ''interactive stands'' allowing users to search for specific items of information by pressing buttons to select objects and their description, thereby conducting their own ''tour'' of exhibits.

Alternative educational, cultural and leisure services

New educational, cultural and leisure services managed by community groups are starting to emerge alongside major public facilities such as museums and media resource centres. These services are aimed at meeting the specific needs of given groups in society rather than those of the general public.

In several countries, microcomputers or videotex are used in adult literacy classes as a way of introducing new technologies into youth clubs or in order to establish links with people living in certain rural areas.

Numerous scientific and cultural clubs and associations also provide opportunities to pursue rewarding educational and cultural activities.

Alternative forms of instruction and training must be developed in order to allow certain groups access to knowledge. It has been discovered, for example, that the use of computers in teaching can help illiterate adults or drop-outs from the school system overcome their fear of institutions such as schools, media resource centres or museums – in many cases they simply cannot afford to buy a computer to use at home. Furthermore, membership of a computer or astronomy club, or a historical society, can help people discover their potential by pursuing a productive cognitive activity outside working hours.

Tying in acquisition and management of knowledge

This rapid overview of the five main areas of educational and cultural development has allowed us to identify the main strands of convergence between sub-systems which until now have remained separate, but which will increasingly have to be seen as an integrated whole.

Public educational services, in-service training, distance learning, basic or alternative cultural and educational resources, are all separate streams flowing from the same source of knowledge and information on which the inhabitants of a city draw.

Within this system, individual citizens routinely produce and consume information – sometimes both at the same time – using various resources and tools at different locations and stages of an overall cognitive process (Figure 1). The peripheral line represents the inhabitants of a city who utilise and/or produce knowledge. They are connected to the "cognosystem" of their city via the cognitive services supplied by the city (areas shown as boxes) and videotex links to other cities. We can now redefine the role of each of the major cognitive services of the future depicted in this figure as follows.

Bodies which integrate and record information

The role of the media resource centre is to integrate knowledge into a coherent system that can then be readily used by other services. A videotex bibliographic search service is available both to organisations that supply and distribute media services and to individual users. This service is supplied through an international network of media resource centres. When required, a media resource centre can supply any of the basic information at its disposal in the form of a videodisk or CD-ROM. These vehicles for raw data can be hired in the same way as the books, films, videos and software in the public collections managed by the media resource centre.

The museum centre plays a comparable role to that of the media resource centre, but differs from the latter in two respects:
- it assembles collections of objects which are then placed on public display;
- the other media it uses focus on a series of specialist themes.

Like the media resource centre, a museum centre acts as a collective cognitive memory to which other services and individuals contribute and on which they can draw.

Bodies providing media services

These organisations transform the first-hand knowledge held by media resource centres and museum centres or supplied directly by specialists. They comprise the electronic and written media (radio, television, newspapers), open universities and universities offering dis-

Figure 1. **The cognitive system of a city**

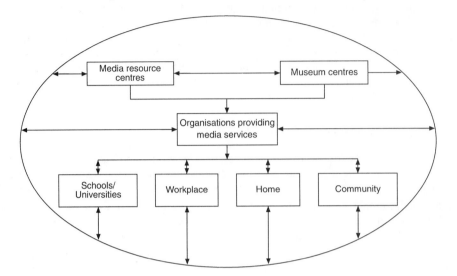

tance learning courses, firms supplying audio-visual material, software and videotex services *etc.* In many cases these organisations are incorporated in disseminating bodies (see below).

Their role is to synthesise knowledge and present it in a more readily accessible format aimed at a variety of user categories. Knowledge processed by these bodies can be incorporated in media resource centres or museum centres, or distributed to schools, the workplace, individual homes or community centres.

Bodies which disseminate knowledge

The primary role of these bodies is to select specific items of knowledge, and the media through which such knowledge will be presented, and then to organise these media into a series of activities that users can participate in directly.

There are four main channels for disseminating knowledge which may be categorised according to where the teaching takes place, the media used, and the type of organisational structure and teaching methods employed:

- schools, universities and other educational institutions which disseminate knowledge in premises earmarked for training;
- the workplace, where specialised vocational training is provided, relating directly to the organisation;
- the home, where information is received in the form of videos, videotex or written texts;
- community centres offering alternative instruction to a specialised clientele as well as introducing to technology those who are unable to use other learning channels.

Institutional arrangements

After this overview of the foundations that have already been laid for the cognitive city of the year 2000, we shall now take a brief look at projects currently in progress in various countries which give a glimpse of what awaits us in the future and which offer tangible evidence of the general approach that has been outlined above.

Public educational services

Planned introduction of microcomputers into schools (France, Canada etc.)

Most countries in Europe and North America have embarked on large-scale programmes to equip their schools with microcomputers. In many cases, however, funding has not been available to make proper use of such resources. Where microcomputers are available, for example, suitable educational software is hard to find; and few teachers have been properly trained in the use of microcomputers as teaching aids.

This situation is fast improving, but two major shortcomings still need to be remedied: funding needs to be found for the development of educational software and learning-activity databanks in a market that is still not profitable for private enterprise; and advisers need to be appointed in each teaching sector to explain to teachers how to introduce new technology into the classroom.

Despite these failings, the overall prospects are good. Administrators, teachers, pupils and parents all agree that computers in schools "are here to stay". Moreover, new types of software are being used for teaching with every year that passes. Success stories include the use of word processing in learning languages, and the use of databases in the natural and human sciences.

Delta project (Europe)

The Delta project (Development of European Learning through Technological Advance) is a R&D programme aimed at the use of new information and communication technologies for initial and continuing education in Europe. In 1989-90 the project received funding of 20 million ecus. The programme was launched in 1988 by the European Community in response to the perception widely felt by all concerned with these issues that the increasingly rapid development of new communications technologies made their use in training an absolute necessity.

"One of the aims of the Delta programme," according to Nicole Duchet, one of its managers, *"is to use advances in technology to improve teaching resources. Such advances also include software and artificial intelligence... Delta must endow Europe with a modern educational system capable of reaching individuals everywhere within society".*

APO-Québec centre (Canada)

The APO-Québec centre was set up in 1985 as part of Quebec's plans to introduce microcomputers into the school environment. It effectively became operational in 1987 with a three-year budget of C$14 million. This money has been used by the centre to form in-house teams and also to fund a number of outside teams carrying out R&D into computer teaching applications. Several of the projects funded by the centre have already borne fruit and the resulting products have been introduced into schools: models and activities for using software packages, systems for designing knowledge-based teaching software, robotics for schools, hyper-media exploration software *etc.*

Videotex and language learning (United Kingdom)

School videotex projects have been carried out in various countries. In the United Kingdom, for example, a number of secondary schools have been given satellite links for use in teaching languages (German, French and Italian). These were designed to establish a videotex link between pupils in the different countries. By discussing current affairs, pupils learn about the culture of the country concerned and the language they are studying. The impact in terms of pupil motivation has been considerable. The use of technology allows pupils to relate a subject learned at school to their everyday life.

Support for vocational training

Vocational training support fund (Australia)

Some countries such as Australia have set up vocational training support funds designed to finance the training needed to keep up with the extremely rapid advances in technology which firms must contend with. The government levies a special tax on firms of 1.5 per cent of their wage bill, which is then redistributed to firms that opt to carry out vocational training at the workplace. Firms benefiting from this provision can recover their contributions in the form of tax relief on training. This measure offers firms an enormous incentive to offer such training, and a simple means of raising considerable sums of money which are then reinvested in society's most precious resource, *i.e.* its human resources and the know-how they embody.

New systems of professional qualifications (United Kingdom)

The United Kingdom has introduced a new system of professional qualifications based on a description by the employer of the skills required for a specific task. Learning is no longer seen as the prerogative of educational institutions. Credits are awarded for previous achievements, notably vocational training, and then recorded in a ''National Record of Vocational Achievement''. The government plans to extend this system of recognition for training credits to encompass, and ultimately to replace, all academic and professional qualifications, which will be integrated into the system with the emphasis placed on skill[11].

Distance education and training

Open University (United Kingdom) and Télé-Université (Canada)

Universities in more than 50 countries now offer distance learning courses. The best-known precursor of distance education is probably the Open University in the United Kingdom. A study[12] of its first eight years of operation, starting in 1972, has shown that a total of 56 500 full-time equivalent students enrolled on courses and 54 per cent of those who enrolled went on to obtain a diploma. The cost per student was less than that of a conventional university.

The Télé-Université du Québec started up a few months after the Open University. In 1990, 15 000 full-time equivalent students enrolled on its courses. They were in the main teachers and administrative staff studying new technology, communications or administration. The university distributes course material by post or over a cable network and includes a prospectus, a manual, a televised series on the Quebec open university channel, and occasionally software. Tutors, each responsible for around 15 students, have one or two meetings with their students during the course and are available for consultation by telephone once a week. The Télé-Université has recently started to use videotex, hyper-media and knowledge-based systems as teaching aids.

Minitel (France)

The Minitel service has enjoyed extraordinary success in France, after its widespread introduction with strong government backing. Hundreds of services are now available in a large number of locations. Minitel terminals are used in the home, at the workplace and in schools, linked by telephone lines to host computers which distribute applications. Educational and cultural applications offered by this videotex system include tourist information terminals in the street (Sète, La Défense). Interactive public terminals in Bordeaux provide free access to certain services, notably course advertisements, cultural activities and library files. The use of videotex in formal education is also increasing, particularly for personalised teaching workshops used in vocational training and distance learning courses. Students can use the terminals to communicate with their tutors or to carry out exercises and tests as part of their courses. Local centres provide students with the equipment they need[13].

Videoway service (Canada)

Cable network services have experienced a growth explosion in Quebec, particularly in the Montreal region. Subscribers can choose from some 40 different channels offering programmes broadcast by the major TV networks, as well as specialised services based on videotex. Since early 1990, the main distribution company has enabled subscribers to rent a new system know as Videoway. A combined microcomputer and channel selector, the Videoway unit allows users to switch from TV mode to videotex mode. In videotex mode, the user selects individual items of information with the remote control unit: news briefs, interactive educational games, lists of cultural activities *etc.* The unit also acts as an interface for interactive TV programmes, allowing users to participate directly in – for example – educational quizzes, to call up a more detailed report on an item summarised in a news broadcast, or to choose specific items of interest in a series of documentary programmes.

Cultural services and facilities

Videodisks and CD-ROM (France)

In France, applications are being found for the storage of knowledge and data on videodisks or CD-ROMs. In Paris, for example, the Musée d'Orsay stores reproductions of works of art on videodisk. These can be viewed at seven locations in the museum, and they serve as a guide for visitors. The La Villette Cité des Sciences has also stored a vast amount of information relating to science and technology on videodisks. The Sainte-Geneviève library uses similar forms of storage media to record information about the medieval manuscripts in its collection, as does the Institut du Monde Arabe.

Medialog and Jean Talon projects (Canada)

The Medialog project is currently being developed in Montreal through collaboration between a research centre (the CCRIT), a private company, the Musée des Beaux-Arts de Montréal and the Cinémathèque Québecoise. The latter two institutions have made part of their collections available in a service which integrates a number of existing technologies. Users will access the service from workstations in municipal libraries. Each station comprises a computer with colour screen, loudspeakers with stereo headphones, a CD-ROM drive, a magnetic-card reader, a laser printer, a VCR, a cassette recorder, and a unit for charging copyright fees. The catalogue of Medialog services will include texts, graphics, photographs, sound recordings and video sequences. Users will be able to explore the contents of the catalogue and reproduce the information obtained via the various media, subject to permission. Users will also be able to establish videotex links with users at other local stations. It will also be possible to reserve time slots on the network for distance learning activities[14].

The Jean Talon project, initiated by the Canadian Studies Directorate of the Secretary of State, aims by 1991 to develop an interactive library containing a large amount of historical and contemporary information on Canada. The library, designed to be used for general educational purposes, will include images, data, sound and text on themes of importance to the development of Canada in the twentieth century[15].

Community centres and services

"Information cottages" (Finland)

"Information cottages" have been set up in a dozen or so rural areas of Finland. These cottages offer accommodation services and contain microcomputers connected by videotex links to databases. The services are designed to familiarise the public with new information technologies[16].

The "Learning Center" (Canada)

The Learning Center in downtown Ottawa has been in operation since 1988. In 1990 approximately 200 people aged from 18 to 76 took part in activities spread over, on average, 11 hours a week. This training centre for illiterate adults makes use of microcomputers and a variety of software programmes to teach users the intricacies of reading, writing and mathematics. It is also a training centre for teachers involved in adult education and provides a test-bench for educational software designed for literacy classes.

Assessment of impacts, limitations and trends

The trends and applications outlined above merit a more detailed investigation than the brief review we have been able to give. We can nonetheless point to some of the impacts they have had, identify some constraints on their use, assess probable implications, and describe the more obvious lines along which they will develop.

Impacts of applications and their integration

The application of new information technologies to education and culture will have incalculable economic and social effects. Awareness is steadily increasing of the need to invest in education and culture.

"We have not yet properly grasped the fact that education lies at the heart of economic growth. The countries which have invested in education are those which are now the most successful"[17].

Educational productivity

The OECD report which was quoted earlier stresses the immense waste of time and effort attributed to the conventional education system:

"Paradoxically, a number of studies suggest that only minutes of each school day are spent on learning-inducing activities, that much time is lost to classroom management activity and to the inability of the teacher to deal directly with every student at the same time".

The way to increase the efficiency and productivity of education lies in changing teaching methods and school organisation to accommodate active learning by the student. The twofold productivity of the student and of the teacher is largely dependent on the technological resources that can now be made available in our schools, but which are still largely absent.

In any case, rapid growth in the amount of knowledge to be imparted to students and the need to train young people in new basic skills have made the traditional approach, in which a teacher teaches the same subject to everybody at the same time, a thing of the past. The rapid emergence of a new education system for a post-industrial society will avoid the enormous wastage of the current education system which is incapable of equipping the new generation with the skills it needs to meet the challenges of the information society.

It will allow teaching to be tailored to individual needs by placing the student within a new context of computer-assisted self-instruction.

Utility and cost of vocational training

We have already drawn attention to the large amounts of money that organisations spend on vocational training. The worst aspect of this is that the funds are primarily used to compensate firms for the hours of working time lost in releasing employees for training and in transporting them to the place where training takes place.

However, rapid advances in technology and the reorganisation of work within constantly evolving organisations are bound to increase the need for vocational training. An effort clearly needs to be made to move towards a concept of vocational training which relates directly to the task in hand and which is provided at the workplace. The assumption on which we should work is that the cost of a training system using knowledge-based hyper-media should not exceed the cost of releasing employees. We should expect such a development to lead to a significant improvement in productivity and product quality.

Optimising cultural resources

The funding that has been poured into major cultural resources such as schools, universities, libraries, museums, audio-visual media and videotex networks has still not yielded the returns that we could legitimately expect. Firstly, some facilities have proved to be ineffective or under-utilised. Secondly, a great deal of effort is wasted when the same work is replicated, more or less at the same time, in the same city and in such equally hermetic contexts as universities, television and museums.

All this work of acquiring, transforming and communicating knowledge needs to be better co-ordinated and integrated. We might then find ourselves in a position where different educational and cultural activities become mutually reinforcing. For example, a hyper-media knowledge base on ecology could be used as part of a course given on a university campus, transmitted into the home by means of interactive television or used as part of an exhibit in a museum devoted to ecology.

The new technologies for representing and processing knowledge facilitate the use of knowledge at several levels and in different ways, while at the same time allowing knowledge to be updated to ensure its continued scientific and cultural relevance.

Development of cultural industries

In European and North American society, cultural activities are becoming the mainstay of economic activity. Expenditure on education and culture must now be viewed as an investment, not only because it increases the efficiency and quality of production but also because it creates jobs in its own right.

Rationalising development

We have to strike the right balance between cultural and educational activities at home, at the workplace and in schools. Whatever the balance finally achieved, it should help rationalise development in both time and space. For example, videotex communications will cut the number of journeys made in the city and will thus help to reduce congestion at peak hours. The amount of space needed to store information will gradually be reduced as paper documents are replaced by digital records. It will be possible to convert large classrooms and use them for more important functions such as science laboratories or multi-media rooms in which small groups can meet.

Constraints

Any change as far-reaching as the one we have outlined above will be accompanied by a gradual but thoroughgoing reorganisation of the entire urban environment. It is only to be expected – especially at the beginning of the process – that certain factors will slow down the spread of the new technologies.

Initial investment costs

The fact that the initial investment is bound to be hugely expensive may slow the process of change until critical mass is attained and services begin to pay for themselves. If we look back at the changes that have taken place over the last ten years, we can see that considerable progress has been made in integrating new technologies into the educational and cultural arena.

At a time when the sophistication of technologies is constantly increasing, rather than resisting change we must actively promote it so that it can be channelled and co-ordinated to greater effect. The time is now right for public authorities to work more closely with other social and economic agencies so that change can be structured, costs kept under control, and efforts concentrated on the most promising developments.

Ignorance of technology

There are many misconceptions about new information technologies. This lack of knowledge fosters wild expectations in some quarters and illogical fears in others. Both reactions impede the spread of the technologies. The fact that such ignorance still exists is another argument for increasing and broadening services aimed at improving the education of the public.

There are still very few specialists in this area, however, and many of them have to teach themselves by using the very methods of assisted self-tuition that they will subsequently pass on to others. Particular attention will have to be paid to ensuring that enough people are trained in the use of technology to meet the demand for educational and cultural services.

Socio-economic, linguistic and national diversity

Different countries, linguistic groups, regions and social classes within individual countries must combat varying handicaps in the race for knowledge. The language of technology is usually English. Therefore measures need to be taken to promote and protect the languages and cultures of other countries.

Economic differences can reinforce disparities between certain groups in society or even between nations. As things stand at present, not everyone can afford to buy a computer or subscribe to videotex services in the home. We must avoid the emergence of a two-tier society in terms of access to knowledge and technology.

Policies must be developed to narrow the differences which currently exist until the critical mass needed to reduce costs has been achieved, as in the case of television. We must make a start on this in the state school system, in which every pupil should have free access to the new cognitive technologies.

In addition to socio-economic inequalities, we must also address socio-cultural differences. For example, some studies have indicated that, generally speaking, women and older people are more uncomfortable about using information technologies. This situation should gradually be remedied through the new anti-sexist values inculcated in schools as well as through the introduction of new technologies at very early stages in the educational process, often at nursery school. Nonetheless, a special effort will have to be made to make the whole adult population computer-literate in the course of the next ten years.

Implications of ICTs and future outlook

It is by no means easy to predict the impact that a given technology will have in years to come. Who, on contemplating the first computer in the mid-1940s, a huge machine packed full of diodes and other highly expensive electronic components, could have foreseen the role that information technology would be playing 50 years later? By the same token, many enthusiastic predictions have come to nought because commentators have failed to take account of the basic features of the technology, the needs of consumers, or the social trends likely to affect the success or failure of the technology.

Technological considerations

We have stressed some of the potential benefits that have prompted organisations, cities and nations to invest in new information technologies. We have also outlined some of the constraints that have slowed down their adoption.

These constraints are currently being overcome, however, through technical advances in terms of hardware and development tools, the development of new interfaces, the introduction of telecommunications networks, and the emergence of new methods of acquiring and modelling knowledge.

The cost of buying a computer has fallen dramatically over the last ten years and should now be within the reach of most organisations. From $100 000 three or four years ago, the cost of a first-rate workstation has now fallen to around $10 000-20 000. Microcomputers at the cheaper end of the market can now be purchased for the price of a TV set, which would seem to bear out Joyce's law that "for equal computing power, the cost of computers is halving every two years".

At the same time, systems are becoming easier to use. The graphics interface concept, with which most systems are now equipped, offers users a far more natural mode of communication. A large number of applications are being developed in the area of speech recognition. One firm, for example, offers an auditive interface capable of recognising 20 000 words in business English. Spoken commands, in freeing us from the time-consuming task of typing in instructions, will make computers even more attractive and accessible to a greater number of people. The possibility of consulting a knowledge base in one's own language will also help to open up access to knowledge for the general public.

Unfortunately, certain applications of new information technologies may require an arduous technical apprenticeship. This forces the user to concentrate on the tool instead of focusing on the area of interest or the problem he is attempting to solve. This situation should rapidly improve, however, with the development of more natural interfaces.

Economic and organisational considerations

The lively interest that the new information and knowledge-based technologies have aroused in economic circles may be attributed not only to their impact on productivity, but also to their potential ability to transform the way in which organisations work.

Major companies have shown faith in this potential, and invested large sums of money in an uncertain area of technology in which much research remained, and indeed still remains, to be carried out. Despite these uncertainties, now that the initial hurdles have been overcome efforts to introduce the technology are proceeding.

The main reason for this is that these technologies meet a vital need for all organisations – an increased ability to process knowledge. In terms of general administration, there is a need to reduce organisational complexity, to integrate an ever-growing volume of information and to consult large numbers of specialists. Operational planning and supervision have to be improved and the operation of complex plant monitored. In the area of financial and support services, employees have to be trained to adapt to rapid change, to introduce new administrative procedures, to circulate new information to the appropriate people, to apply a more intelligent form of financial control, to monitor the development of human and technological resources, and so on. Set against a background of constant change, all these tasks call for the use of appropriate tools for acquiring, processing and communicating knowledge.

The adoption by organisations of second-wave information technologies will bring radical changes in the way they operate, much as management software did in the 1970s and office automation in the 1980s. It will enable organisations to meet the needs of post-industrial society for an organisational environment of higher quality than the one we have now.

Consensual management, one of the driving forces behind the success of Japanese manufacturing industries, marks a clean break with the Taylorism of the industrial revolution. Workers on Ford's first assembly lines were not required to think about what they were doing; their actions were repetitive and pared down to what was necessary; they were classified into rigidly defined categories and their better-paid supervisors were responsible for quality control. In contrast, consensual management is designed to develop and retain a labour force that is skilled, versatile, co-ordinated and committed, to streamline management, to give priority to innovation, to foster the sense of partnership between management and the unions, to reduce the number of job categories while broadening their compass, to provide incentives for high-quality work and innovation, and to update recruitment and training methods.

Work and cultural values

The new information technologies will have a profound and far-reaching impact on cultural values and the way in which people work. A study of 24 firms in the United States which had started to introduce new management procedures revealed an increase in the number of skills required of workers. Operators had a number of functions, some of which were formerly carried out by technicians, and technicians were now doing some of the work that used to be given to engineers. Supervisors are becoming "fixers" rather than leaders. The responsibilities of management have been both increased and broadened.

There has also been a significant increase in the intellectual content of jobs. Until now, management software and office automation have simply freed employees from the more mechanical and routine tasks: file searches, compilation of statistical data, text formatting, *etc.* The new knowledge-based technologies can go even further by automating higher-level tasks.

By and large, jobs will become more rewarding but also far more demanding. A number of workers will probably be unable to keep pace with these changes. Just as industrial development deskilled physical work which is now largely taken care of by machines, knowledge-based technologies will have the same impact on tasks with a smaller intellectual content. Before computers came on the scene, ability to calculate was highly prized. Now that calculations can be done by machines at unbelievably high speeds, such tasks will no longer be taken into account in assessing the suitability of job applicants.

The development on a massive scale of software programmes capable of performing very large numbers of intellectual tasks will also result in greater emphasis being placed on the attributes that distinguish man from machines: common sense, the ability to pick out relevant information from a welter of facts and figures, the ability to learn and to make judgments. In this way, knowledge-based technologies will gradually bring about changes in our values and the order in which we rank them. Imagination, judgment, creativity and innovation will increasingly become the bywords of tomorrow's world.

The socio-political setting

The changes wrought by the emergence of an information society, and in particular by the development of increasingly intelligent computer tools, are so far-reaching that opinions about these changes are frequently radically opposed. Some commentators have enthusiastically welcomed the advent of intelligent machines, believing that by helping us to solve the serious environmental, economic and social problems that we are facing they will allow us to transform the world, while others are pessimistic about the future of our society, fearing that humans, displaced by machines, will have no part in it.

The widespread use of new knowledge-based technologies may well make people less dependent on organisations and the specialists they employ such as doctors, architects and lawyers. They will supply the consumers of services with information which will enable them to ask the right questions of the experts they consult. In medicine, the new technologies will help to improve standards of health in developing countries by making it possible to give high-quality, low-cost training to paramedics providing services in large geographical areas. They can also help us to solve our environmental problems and to use the Earth's resources more rationally.

The other side of the coin is that we may well find ourselves spending our days signing forms churned out by these expert systems without having the faintest hope of knowing exactly what we are committing ourselves to. Furthermore, the adoption of new knowledge-based technologies raises highly complex moral and legal issues. For example, where does the responsibility lie for medical decisions made with the help of expert systems? With the doctor, the experts who provided specialist input for the system, or the company marketing it? These issues, and others besides them, call for a painstaking revision of our laws and regulations.

In some respects, we should have nothing to fear from the fact that machines might prove to be more useful than human beings. Man, in whose image we attempt to make machines, will doubtless prove to be an elusive, chameleon-like quarry. Indeed, by enriching the cognitive environment, the new information technologies will allow mankind to progress even further:

"human beings constantly exploring new frontiers, and transmitting to the machines the more stable elements of their knowledge and methods"[18].

There is a consensus among commentators and researchers involved in artificial intelligence that the new knowledge-based technologies, like any other technology, have considerable potential both for good and for harm. Echoing a sentiment voiced by one of the pioneers of artificial intelligence, John MacCarthy, we might perhaps say that if there are human tasks for which a computer cannot be programmed, such tasks should not be carried out by anybody or by anything, be it a human being, a system or a machine. Examples that spring to mind here are the use of expert systems for military purposes and of telecommunications for spying on people.

To avoid abuses of the second wave of information technologies, we should encourage the creation of such organisations as "Computer Professionals for Social Responsibility" in the United States. We should also introduce more democratic procedures into the choice of technology by governments, cities and organisations responsible for knowledge processing. Lastly, public bodies need to be set up in all countries to conduct research to help formulate policy, support socio-economic dialogue, disseminate information and encourage the public to learn more about technology.

RECOMMENDATIONS

In rounding off this admittedly somewhat cursory review, we may conclude that new information technologies are by no means a temporary or minor phenomenon. They are part of a groundswell whose force must be harnessed without delay. A flexible package of measures, co-ordinated at national, municipal and organisational level, must be introduced over the next ten years in order to achieve the cognitive city of the year 2000. In view of the enormity of this task, our recommendations must necessarily be of a general nature. The degree to which they are implemented will depend on a variety of factors determined by the specific requirements of individual cities and countries.

Assessment, planning and discussion

The introduction of new information technologies raises the major social issue of how to avoid creating two classes of citizen, the cognitive "haves" and the cognitive "have-nots". Furthermore, the number of agencies involved and the scale of the tasks they face call for broad-based socio-economic dialogue and new back-up resources. We therefore propose that:
- Local authorities should set up a standing review body for educational and cultural planning under their jurisdiction. This agency should comprise managers, producers and users from a variety of educational and cultural networks: schools, universities, professional training agencies, media and videotex agencies, libraries, museums, community training departments, manufacturing firms.
- A new technology search and evaluation agency should be set up alongside the above body to carry out the surveys and searches required for ongoing assessment of the benefits that the agency and the population may be able to derive from new technologies.
- The agency should organise regular public debates to formulate action plans, and these debates should be broadcast through the media.
- On the basis of recommendations made by the agency, each city, in collaboration with central government, should gradually revise its budget and its regulations in order to keep pace with the installation of new information and knowledge-based technologies.

Development of public educational services

The foundation for the city of the year 2000 has to be laid in schools. The state school system is in the best position to ensure that all social, economic and cultural groups are given fair access to the new technologies. We therefore propose that:

- primary school curricula should be adapted to provide the new basic skills required, and should focus on knowledge processing and active learning by pupils; a system of knowledge processing workshops should be introduced at other levels of schooling;
- subsidies should be provided for the development by enterprises or research centres of "model" knowledge-based training systems which would allow the introduction of teaching methods better suited to the needs of the information society;
- teachers should be given access to suitable computer tools with links to media resource centres, museums, educational TV broadcasts and videotex services;
- new technology advisers should be appointed in schools to advise teachers on how to introduce new technologies;
- initial and in-service teacher training should be geared to the basic skills and learning facilities offered by new information technologies;
- the educational system should be brought into the post-industrial world by gradually freeing it from the rigid regulatory constraints of the industrial society that currently dictate its *modus operandi*.

Support for training by firms

The ongoing rapid development of knowledge-based systems means that firms must train employees for specific tasks. The prime concern of firms must become the quality of their human resources and the knowledge embodied in these resources. We consequently propose that:

- A technology and training development fund should be set up in collaboration with central government, to be financed through a special tax on companies' wage bills. A firm would be entitled to a tax credit varying according to the amount of R&D and training it carried out.
- Central government should be persuaded to set up a new national system of training certification that would take account of the diversity of resources and places of training, and recognise in-service training through training credits.
- Part of the development fund's resources should be used to provide financial support for the development of knowledge-based hyper-media training systems applicable to the operations of a large number of firms; these systems should be made available at low cost in order to promote their rapid spread and development by firms themselves.
- Co-operation should be encouraged, either through financial incentives or by regulations, between educational and cultural networks (notably universities) and firms, to ensure that equipment and human resources are used to best effect.

Distance learning

Distance learning not only affords certain categories of the population access to knowledge, it also allows individuals to tailor courses to their requirements in terms of when and where they study. We therefore propose that:

- Central government should be persuaded to set up a Distance Learning University designed to promote high-quality training at home, at the workplace or in collaboration with community services; this university should also be made responsible for the accreditation of training in the context of the newly introduced national system of certification.
- In collaboration with central government, a larger budget appropriation should be made for the development of communication networks, particularly cable networks, educational and cultural videotex services in the home, and remote access to media resource centres and museum facilities. The development of these various services should be co-ordinated.

Community educational, cultural and recreational services

Public services and facilities must be brought into the neighbourhood. Community-sponsored educational and recreational services provide an invaluable foil to other services and as such deserve both recognition and financial support. We therefore propose that:

- action should be taken to encourage media resource centres and museum centres to make greater use of the different media and to make the knowledge in their possession available not only on their premises, but also in schools, at the workplace, in the home, and through community educational and cultural services;
- local cultural and information centres should be set up in each neighbourhood of major cities and in rural areas to allow the public first to obtain a grounding in information technologies and then to learn how to use them to gain access to the knowledge available on various educational and cultural networks;
- a programme of support for community agencies promoting the use of new technologies should be set up, applying the principle of ''positive discrimination''; such agencies would be in the best position to offer literacy classes, to establish contact with school drop-outs and to raise awareness of the applications of technology among certain social classes and groups, notably the elderly.

NOTES AND REFERENCES

1. GODIN, S. (1989), *L'actualité,* Montreal, October.

2. DOYLE, D.P. (1989), *Business Week.*

3. CENTRE FOR EDUCATIONAL RESEARCH AND INNOVATION (CERI) (1986), ''Information Technologies and Basic Learning'', General report, OECD International Conference, Paris.

4. CONSEIL SUPERIEUR DE L'EDUCATION (CSE) (1988), ''Rapport annuel 1987-88 sur l'état et les besoins de l'éducation. Le rapport Parent, vingt-cinq années après'', Québec.

5. GROUPE REPARTIR (1990), ''L'école de demain et les nouvelles technologies de l'information'', *Revue Le bus,* special issue, Montreal, February.

6. DOYLE, D.P. (1990), ''Endangered Species: Children of Promise'', in *Business Week,* Special Bonus Issue, McGraw-Hill, pp. E1-E136.

7. FRENETTE, J.G. (1990), ''Projet PRESENS, rapport d'étapes'', Rapport interne au CSS Lanaudières, March.

8. HENRY, F. and KAYE A. (1985), ''Le savoir à domicile'', Presse de l'Université du Québec, p. 330.

9. DANILOV, V.J. (1982), ''Science and Technology Centers'', MIT Press, p. 42.

10. FINAL REPORT BY THE COMMITTEE (1991), ''Le centre d'interprétation de l'histoire de Montréal'', March.

11. ENNALS, R. ''Knowledge-Based Systems and Work-Based Learning'', *Journal of Artificial Intelligence in Education,* Vol. 1, N° 2, p. 57.

12. MCINTOSH, N., WOODLEYR, A. and MORRISSON, V. (1980), ''Student Demand and Progress at the Open University, the First Eight Years'', *Distance Education,* Vol. 1, N° 1, pp. 37-60.

13. OECD (1990), Project on Urban Impacts of Technological and Socio-demographic Change, National Contribution from France, URBA 2000, pp. 8 and 19.

14. CENTRE CANADIEN DE RECHERCHES SUR L'INFORMATISATION DU TRAVAIL (CCRIT) (1990), ''MEDIALOG, un service d'accès aux multimédias'', project summary.

15. OECD (1990), Project on Urban Impacts of Technological and Socio-demographic Change, National Contribution from Canada, pp. 15 and 16.

16. OECD (1990), Project on Urban Impacts of Technological and Socio-demographic Change, National Contribution from Finland, p. 29.

17. GODIN, S. (1989) *op. cit.,* p. 22.

18. PAQUETTE, G. and ROY, L. (1990) ''Les systèmes à base de connaissances'', Télé-Université and Editions Beauchemin, Montreal.

BIBLIOGRAPHY

BERGERON, A. and PAQUETTE, G. (1990), *Discovery Environments and Intelligent Learning Tools,* in Gauthier G., Frasson C. (eds.), *Intelligent Tutoring Systems: at the Crossroads of Artificial Intelligence and Education,* Ablex.

BORDIER, J. and PAQUETTE, G. (1990), *Building Learning Environment Using Generic Software,* communication selected at the 4th World Conference on Computers and Education, July, Sydney, Australia.

PAQUETTE, G. (1987a), *L'apprentissage des fractions à l'aide de logiciels outils,* Actes du colloque de l'AQUOPS, April.

PAQUETTE, G. (1987b), *L'activité scientifique à l'école, des progiciels d'application aux logiciels-outils intelligents,* Actes du colloque CIPTE 87, J-Y Lescop, éditeur.

PAQUETTE, G. (1987c), *L'intelligence artificielle et l'éducation,* Bulletin de l'AQUOPS, May.

PAQUETTE, G. (1988), *Le dévelopement d'outils intelligents d'apprentissage pour le traitement des connaissances,* Actes de ITS-88, Université de Montréal, June.

PAQUETTE, G. (1990a), *L'acquisition des connaissances sous forme de règles pour l'apprentissage,* Revue ICO, March.

PAQUETTE, G. (1990b), *Discovery Tools for Rule-based Knowledge Learning,* communication selected at the Conference for Advanced Research on Computers and Education, July, Tokyo, Japan.

PAQUETTE, G. and BERGERON, A. (1990), *L'intelligence artificielle,* Télé-Université and Editions Beauchemin, Montreal.

PAQUETTE, G. and ROY, L. (1990), *"Les systèmes à base de connaissances",* Télé-Université and Editions Beauchemin, Montreal.

NEW INFORMATION TECHNOLOGIES
AND UTILITY MANAGEMENT

Gabriel DUPUY
Professor at the Institut d'Urbanisme de l'Université de Paris
Professor at the Ecole nationale des Ponts et Chaussées
FRANCE

Gabriel Dupuy is an engineer from France's Central School of Arts and Manu-facturing, with a doctorate in Applied Mathematics. He also holds a doctorate in Letters and Human Sciences, a doctorate in Urban Sociology and a diploma of Higher Studies in Economics.

In 1984, he won the Jean-Jacques BERGER prize awarded by the Académie des Sciences and, in 1989, the Abel Worman prize awarded by the American Histor-ical Society of Public Works.

A professor at the Institute of City Planning of the University of Paris XII, he also teaches at the Ecole nationale des Ponts et Chaussées and at the Engineer-ing School of the City of Paris. In addition to his teaching, Gabriel DUPUY is engaged in research, notably in association with the laboratory "Techniques, Territories, and Societies" of the Ecole nationale des Ponts et Chaussées. He directs the "Networks" research group of the National Centre for Scientific Research.

He has collaborated on several volumes of analysis of urban topics and pub-lished numerous articles and papers.

New information technologies are making spectacular progess in most fields of activity in modern society. The management of urban transport and communications networks, energy and other systems is no exception. Every OECD country now has several years' experience of such applications and is introducing current innovations as well as planning for the future. What is needed, however, is the identification of all the specific features of this sector, differentiating it from industry or public management for instance. This is the subject of the first section in this paper. Next are a review and examples of the major applications known to date, with particular emphasis on the reasons why new information technologies have been used in each case. The third and final section attempts to identify current implications and future trends in the field of urban network management. A number of recommendations conclude the paper.

FEATURES OF THE SECTOR

The specific way in which new information technologies are used in urban networks can be better understood by looking at the three terms "network", "urban" and "information" as they are understood today.

Networks today

Networks, or more precisely technical or infrastructure networks, are not a recent invention. It was mainly in the last century that cities began to instal transport systems and lay water, sewage and gas pipes, telegraph and telephone lines, electricity cables, *etc*. The trend has continued up to the present day, with the adjustment of the road system to car traffic and the introduction of new networks such as cable television.

Within the space of one and a half centuries, these networks have taken on a specially important role. Originally designed for quite a small number of users, often on the basis of supplier/customer contracts, they have since grown considerably in number and in size. Over the years, town-dwellers have been connected up to more and more networks (public highways, public transport, water, electricity, gas, sewage, urban heating, telephone and so on). At the same time, the number of consumers has gradually been increasing. Today, major cities in industrialised countries often boast connection rates of 80 to 90 per cent or more. Access to the various urban networks is no longer an "extra" or restricted to certain districts or to certain social groups. It is looked upon as an absolute necessity, a social right, perhaps the very symbol of urban life. Not only do networks provide the consumer with cubic metres of water, kilowatts of electricity or quantities of teledata. They also link him up to the modern world which is also involved in the supply and demand process for water, electricity and communications.

As a result, access to the networks and their efficient operation have become vital issues. Unlike earlier times, there is now a tendency to overrate networks and take them for granted as a basic element of personal comfort, and a prerequisite for social integration. The outcry caused by proposed network rationalisations whenever they entail the prospect of certain population groups being deprived of services or refused access is proof enough of this tendency.

In the United States, the right to be connected to a network, in this case the telephone, was set down through the Lifeline system. Deregulation in the telecommunications sector had called into question the time-honoured system involving cross-subsidisation of household calls by corporate communications and of local calls by trunk calls. The ensuing rise in local call charges to households had meant that those in difficulties (due to social problems, unemployment, *etc*.) were unable to pay their bills. Their telephones were then disconnected by the telephone companies. Civic rights associations immediately fought to have those who had been "cut off" reconnected and prevent any new cases occurring. After a bitter struggle, the right of every individual to a guaranteed minimum local service was recognised and given the name "Lifeline". The main argument was that, in modern American towns, the very existence of any citizen, especially the socially and economically underprivileged, depended on the possibility of using the telephone to ask for assistance, call for emergency or medical help, look for a job and so on.

"Lifeline", like its name, is an example that tends to speak for itself. But it is by no means the only one. In California, utilities intending to deregulate their prices have recognised the consumer's right to a guaranteed minimum supply of electricity. In other countries, basic supply (of water, gas, electricity and telephone services) is guaranteed, even for those who cannot afford the connection charge.

All this must be taken on board by network managers for there is little likelihood of the situation being reversed. Basic network services must be widely extended to include all those who want access. Service must be of consistently good quality, since it has become highly symbolic (just as bread used to be, in some European countries). Pressurised water supply, instant sewage disposal, regular refuse collection and continuous electricity supply are not luxuries but standards of service which, if not met, lead to protest.

So whether there is a tradition of public service or not, network operators must provide everyone with a reliable, high-quality service. This raises some thorny problems for managers, obliged to supply services at minimum cost (so that everyone has access) but at a standard that ensures there will be no peak-time or supply hitches.

"Urban" issues today

While "urban network" is a common term, the word "urban" is increasingly difficult to define. Recent times, as we know, have been marked by growth in the size of cities, usually measured by the number of inhabitants. But today, what is changing is not so much town size as urban patterns. The difference between urban and rural environments can no longer be used to trace city boundaries. In the past, cities grew up on the principle of contiguousness, *i.e.* immediate physical proximity of buildings. Today, however, analysts are increasingly aware that, while the term "urban system" is still appropriate, it is less a question of contiguousness or vicinity than of access, mainly through the networks mentioned above. Because networks enable people to make contact and offer a range of links and transactions, they allow urban systems to operate without being excessively tied to the notion of contiguousness (or, for that matter, the idea of a central point or at least a single central point).

Networks were originally designed to serve the most densely populated parts of urban areas. Building and operating standards were drawn up to suit continuously built-up areas, or even density requirements, all of which is increasingly out of phase with the cities of today. The problem now facing network managers is to operate (entirely separate) sets of lines, pipes and small networks scattered over a considerable area and then interconnect them all in order to benefit from economies of scale and network effects. Time has been a deciding factor with some networks, for which regional or national scale operations have to some extent proved to be the solution. This is true of the telephone system and electricity supply[1]. But the problem is still acute for water and sewage systems, public transport and household refuse collection, not to mention road traffic management.

The "information revolution" and "new information technologies"

Some explanation must be given at this point with regard to information and the new technologies that facilitate data transmission and processing. It is tempting to see current developments merely as a sign of very rapid technological change. Yet the "information revolution" is also, perhaps above all, something very different. Its outcome is the application and acceptance, in all fields, of the concept of information as knowledge presented in such a way that it can be understood and acted upon. Not confined only to computers and fibre optics, a system has grown up to define encoded knowledge that can be transferred or stored, and today that system can be used to create networks. The distinction, which is not purely theoretical, has its relevance here. Firstly, network management has always involved the transmission and processing of information. After all, electric dispatching systems were used for the first time by INSULL in Chicago at the turn of the century. The first urban telegraphs, dating back to the 1850s, were mainly used as alarm systems (to call the police or fire brigade). So new information technologies have not played a truly revolutionary role in network management. Information, in the modern sense of the word, is chiefly a means of accelerating processes and rationalising options, but above all it can now be handled, stored and transferred much more easily than before.

Before any further description of the part played in network development by the information "revolution", it should be remembered that a generation earlier there had already been a "quiet revolution", namely that of the telephone. Telephone technology has gradually become economical and reliable, and hence an integral part of management for transport, energy and other networks. Using dedicated telephone lines or the public system, staff have been able to contact one another and their network control stations. "On-call" systems, allowing staff at home to be called in to work at night or during holiday periods if an incident should arise on the network, had already become widespread well before the emergence of new information technologies. For a better understanding of the situation today facing those who run urban networks, it is interesting to look back at the comparison drawn by Peter Hall at a meeting of OECD experts in 1987[2]. Comparing the development of the automobile system with that of information technologies, Hall pointed out that the basic technology needed for the automobile was already available by 1910. We are all familiar with what followed, *i.e.* the tremendous development of an automobile-based system, but it was due not so much to a technological revolution as to the fact that a genuine system had been set up, including the mass-production of cars and the development of a series of car-servicing centres, a set of traffic regulations, a system of paved roads and, more recently, a network of hotels, restaurants *etc.* catering to the automobile traveller. Hall pointed out that one of the basic components of "New IT" was the telephone, invented in 1876, somewhat earlier than the car (1885). The history of the telephone is similar to that of the automobile-based system. But in Hall's view IT development has gone no further than the car had in 1916. New information technologies, still hardly more advanced than the telephone, resemble the Model T Ford in the 1920s – mass-produced, driven by largely novice drivers on primitive roads that end at the edge of the city. No developed infrastructure, no overall system (like the automobile-based "system") has grown up around new information technologies. This is the avenue that Hall invited us to explore, while stressing that relatively swifter change in this sector might telescope the process. As far as urban network management is concerned, it is this path that holds the greatest potential for significant change and promise for the future, and not some individual, one-off innovation that owes nothing to the advances, however significant, already made in telephone technology.

The sphere of urban management

Within the broad context described above, characteristic trends in urban management are emerging in many countries. The trends are important in that they shape network management policy and are therefore worth reviewing.

The first noticeable trend is towards handing over the management of local affairs to those directly responsible. It is now an accepted fact that the rational use of resources becomes extremely complex beyond a certain point. In countries where central government used to manage cities right down to the slightest detail, decentralisation is devolving power back to local authorities; this is the case in France, New Zealand and Turkey, for instance. In others, such as the United States or Italy, central government is making lower level territorial authorities responsible for local matters. In many other cases, management of some urban services is being handed over to private enterprise, which becomes financially responsible for all or part of the operations. The trend is very marked throughout the world, for instance in the United States, the United Kingdom, Germany and France. It is particularly noticeable in the case of services provided by urban networks, some of which had previously, for historical reasons, been run directly by the public authorities (water supply and sewage systems in particular).

To some extent, this overriding trend is bringing network management closer towards corporate behaviour and further away from previous, more bureaucratic practices. Nevertheless, it must be stressed that the inflexibility originally found at organisational level has remained unchanged throughout these developments. There has been no revolution in the way urban networks are organised to compare with the one in industrial production. This stability may be partly due to inertia and change may occur over the coming years. But this is a very specific sector that should not be underestimated. Networks remain territorial and the public service factor mentioned earlier still prevails as does the link with local policy, even though management methods are evolving. As a result, the initial organisational arrangements, relatively well suited to all these factors, tend to remain unchanged.

A second important trend is related to the increasing emergence of social problems. Without probing too deeply here into sociological factors which go beyond the scope of urban network management, there are four major aspects to be considered:

- *Individualism.* People everywhere are increasingly attached to values such as independence, choice and freedom that cater first and foremost to the individual and personal preferences, "one-to-one" treatment, *etc.* This trend conflicts with "mass" management and systematic standards, which have until now tended to prevail in urban network operations.
- *The minorities issue.* As social segregation increases, there is growing sensitivity to the precarious or even tragic situations facing some social groups or individuals. In industrialised countries, urban society is no longer willing to be openly confronted with areas of dire poverty, total deprivation and hopelessness. For the reasons mentioned earlier, urban network management in particular should not ignore this trend, even though it conflicts with other constraints such as cost-effectiveness.
- *Environmental protection.* This is merely mentioned in passing. Networks are often involved here, whether directly when water supply and sewage systems deteriorate natural resources, or indirectly when the building or use of infrastructure harms its surroundings. Urban environments are particularly sensitive and operators investing in and running their networks must work under considerable constraints that affect com-

pletion dates, costs and even management procedures (communication, consultation with users or residents).

– *Security.* Social change explains the dramatic increase in the demand for security, particularly in cities. For instance, citizens no longer feel protected by the web of solidarity that once existed between relatives, neighbours and colleagues, sometimes all living in close-knit communities. Nowadays, people look to other quarters – special services, government, *etc.* for ensuring the security they see as essential to life in the city. It could be said that such demands are made on every urban network because it provides a link with the outside world (see above). But new information technologies with their potential for rapid transmission and interaction are increasingly becoming economic substitutes for earlier forms of security and protection. So network managers too are directly concerned by security issues of every kind.

Major applications for new information technologies in urban network management

Scope

The survey carried out in France by URBA 2000 showed that the transport sector and the utilities combine to form the main area of use for new information technologies[3]. In fact these applications could be used in any type of urban network. To date, however, the problem of conveying people from one place to another does appear to have attracted the most "solutions", to use a term from the OECD questionnaire sent out to Member countries[4]. This is hardly surprising. A combination of many factors are inducing operators to adopt new information technologies: new urban patterns, now much more fragmented than before; the need to ensure that everyone has the right to travel in economically acceptable conditions on a fair basis; a much-needed change in operating practices, still catering more for mass traffic than for widely differing uses, quality of service and user choice. New IT applications have made particular inroads into fields such as conventional traffic control (traffic lights), vehicle guidance and parking and have become public transport management tools for operating routes and providing travel information. There are also applications relating to tickets and payment, both in individual transport (automatic tolls and fines) and public transport ("smart cards" that allow differentiated pricing, involve no extra staff and are user-friendly).

Although transport, a specific case among urban networks, is a sector worth highlighting and relevant to this study, it will not be covered in greater depth here since it features in another part of this book.

For similar reasons, we will not be giving an in-depth analysis of the wider environmental applications (particularly air pollution) or energy applications of new information technologies. Where the environment is concerned, problems are usually not those encountered in running networks. The environment is such a vast subject that the limited context of networks may even be a stumbling block. Since environmental applications are also covered in another section, this paper will study only urban water supply and sewage systems. As for energy, apart from the transport sector, the main area affected by energy-use monitoring is housing. However, with the exception of urban heating systems (mostly in Scandinavian countries and Germany), this type of monitoring is less a question of network management than of heating techniques. This study will therefore be restricted to the new IT applications in electricity, gas or urban heating networks, and will not touch upon the wider, special issue of energy use.

The networks covered in this paper will therefore include:

- water supply;
- sewage systems (rainwater and wastewater);
- electricity supply;
- public lighting; and
- solid-refuse collection.

To these should be added security systems generally based on information transmission and processing.

With the exception of the last on the list, each sector is distinctly industrial in nature. Production and management conditions are not those prevailing in industry, as we have seen, but user relations are similar to those linking any firm and its customers. Here, unlike the transport sector, users are only marginally involved in the production of goods or services provided by the networks[5]. So, heading the list of applications used in the networks come the new information technologies aimed at internally rationalising production, achieving more efficient use of resources and a better match between supply and demand. This will be the first type of application described below.

But precisely due to the nature of the information that new technologies help to collate, send and process, networks no longer have to be managed on a strictly individual basis. A number of applications specialise in the cross-network exchange of information that may help in managing one network in relation to another, or to a broader environment (town-planning specifications, for instance). These applications will be presented under a second heading.

Finally, security applications, insofar as they are not concerned with merely rationalising the administrative side of police or fire departments, are a special case.

Rationalising urban network operations

Providing users with services through a network means, in fact, putting production/processing units into contact with consumer units, *i.e.* users (or user installations), in time and in space and as efficiently as possible. It involves information, provision of the required services, invoicing, *etc.*

In an ideal system, network operators would have "real-time" control over all their production and consumer units, but network operations as they exist today run into numerous obstacles. Production and use occur at different points, use varies over time, some production/processing units or sections of the network may be unavailable, communication with users may pose problems, and a variety of potential hitches may adversely affect supply in regard to demand.

So for some time now network operators have been using a series of planning schedules that they have developed to suit different management time-horizons. Over the long term (several years), investment decisions based on approximate demand/consumption forecasts rationalise production and distribution. Over the shorter term, *e.g.* on an annual basis, planning schedules allow for given production/processing units to come into existence, increase output or connect up to the network in a given time or place to cope with specific demand. Finally, over the very short term (varying from a few hours to real time), operators attempt to plan the adjustment of network operations and ensure that they can cope with hitches, failures, unexpected peaks, *etc.* Different networks naturally adopt different time-scales but the principle is broadly the same, and any advances in the transmission and processing of information will affect these three types of planning.

In long-term planning, increasing use is made of modelling techniques. Designed to optimise system design, investment and so on, models are becoming more practical thanks to:

- Increasingly extensive computerised databanks (operating data on previous years, costs, etc.).
- Much higher data-processing capacity, making possible the comparison of numerous variables. This is a most important point. Any change in a network can produce a combination which is impossible to process manually or at sufficient speed even with early-generation computers.

One illustration of how these applications are used in various countries is electricity supply networks.

Medium-term planning has derived even more benefit from advances in computing. The case studies in the OECD survey give some good examples of this kind of application. Canada has described how household refuse collection has been optimised. France has also developed similar applications. This has been possible thanks to the creation of extensive, updated and reliable databanks containing information on routes, timetables and where the various quantities were collected. It is worth noting that for an investment of only FF 25 000, an operator in Nantes (France) has fitted a refuse lorry with sensors to collect and record this type of information on a cartridge later read by a microcomputer running the databank. This has so far been only a small-scale experiment, of course, but the whole fleet is soon to be fitted with the system, thereby providing considerably better data and optimising refuse collection planning. Many examples are to be found in the field of maintenance. Databases are used to plan repair work on sections of the network and time the ensuing unavailability of such sections. In some cases, economic optimisation becomes possible. This is true of the electricity-supply and public-lighting networks in various countries. Water-supply and sewage systems are a slightly different matter. With new information technologies, in particular through the routine use of sensors to record information for subsequent transfer to databases, operating schedules can now be drawn up (for drinking-water production or wastewater processing cycles). But in most cases, it is not enough to be familiar with the system, parts of which are very old, for preventive maintenance to be rationally planned. Specific applications do exist, such as the use of video cameras to inspect sewage pipes. First introduced in the United Kingdom some ten years ago, they are now in wide use. In any case, medium-term planning for network operations should include a link-up with a geographical database that determines precisely where and how to carry out any necessary work (see para. below).

Other examples include production planning methods, setting the output levels needed to supply a network in a given area at a given time. Electricity supply and major urban heating networks have been or are being equipped with this type of system, in which computerisation replaces manual methods of data collection and processing. For the time being, apparently, this does not apply to the same extent to water/sewage systems, except for the running of production, storage or drainage equipment (drinking water treatment, rainwater reservoirs or storm drains).

The third and probably most promising case is that of very short-term planning, sometimes incorrectly described as "real time". Broadly speaking, there are two types of planning, one being local control and the other central control. In local network control, automatic systems using microprocessors and occasionally microcomputers react to variations in the external environment, in the use of the service or in consumption by adjusting supply flows. But in central control systems (often called centralised management), locally obtained information is sent back to a central station where it is processed and triggers swift adjustments to the system by remote control.

In central control systems, wide use is made of teletransmission. Furthermore, the central unit must be equipped with computer units suitable for high-speed processing, at-a-glance visualisation, *etc.* Local control obviously enables only limited intervention on the network. But then again, central control can only work if it is not cluttered with local information. So networks are often run using a combination of local control and centralised management. Rainwater drainage networks are a typical example. Some years ago, they were managed solely on a long-term planning basis, the size of wastepipes being based on maximum rainwater level forecasts. Since then, the objective has become optimal network capacity, using real-time management techniques during wet weather. Local control systems open flood-gates to drain water held in temporary storage areas. But for a network covering a wide area, centralised control makes it possible to even out rainwater distribution with the help of radar images supplied several minutes earlier. Similar applications, already operating in the United States, Germany and France, are becoming more common.

One example in France is Seine-Saint-Denis. This is a vast area covering a whole département in the Paris area. The flat terrain makes rainwater drainage a problem, and outlets are situated on the far edges of the area. The sewage system was originally web-shaped. The departmental authorities consequently thought of using the web of sewers and adding a number of reservoirs to facilitate the temporary storage of water during heavy storms. Remote control devices were fitted to floodgates, enabling certain sewers to be shut off or reservoirs closed; pluviometers were installed throughout the areas and the system equipped with sensors. After an initial attempt at entirely centralised management, which swamped the control station with information, it was decided to combine local adjustments and a central control unit. The system is run by developing a comprehensive management strategy based on rainfall levels and co-ordinating the various local-control microcomputers to prevent flooding, minimise pollution during wet weather, ensure management in dry weather and improve safety for those working in the sewage network. The system is based on a set of 160 local remote-controlled monitoring stations. Eighty per cent of them generate information and trigger alarms, while the remaining 20 per cent make adjustments to the system on a local basis using adaptable sets of instructions. The system includes animated display panels and indicators for an interactive dialogue with the operator. It can now ensure a form of management that adjusts to highly variable amounts of rainfall. This is not "real time management" in the true sense of the term, but system response lags are very short (a matter of minutes), despite the high number of calculations needed to optimise the exploitation. The system was recently improved with the addition of a radar device placed upstream from rainfall routes to detect in advance how the rain is likely to fall. This gives a time-lapse for anticipatory action that makes it comparable to a "real-time" system.

Water-supply systems rarely use this type of application because they often contain permanent storage reservoirs that act as a ready-made adjusting mechanism. Any applications that do exist, such as the one in Besançon (France), are confined to water control in factories, pumping stations and large reservoirs but not user supply. They have however improved the management of water loss and energy use for pumping. Istanbul (Turkey) has a remote network-monitoring system that can detect faults, diagnose problems and make repair-work decisions. However, water system managers are also aiming to set up very short-term local and central controls designed to offset major quality shortcomings. Sensors will be the main solution, but they must be numerous, running right through the network, efficient (so as not to let any pollution go unnoticed) and reliable.

Centralised management by remote control is very common in urban heating networks. It may consist of monitoring and failure-detection systems or, in the larger networks, genuine regulation systems to match heat supply to consumer demand. Systems combining local automation and centralised control have also been set up to run public lighting systems in various countries (switching on and off, detecting malfunctions).

Network management is not solely confined to the distribution of products or services. It also includes managing user relations, particularly from an economic standpoint. Everything related to meters, consumption and invoicing is of the utmost importance, all the more so when the consumer demands clear invoices and protests against unwarranted inclusive-charge billing. This is a genuine problem, in that the central management unit must somehow be linked to each one of its consumers, all of whom are by definition scattered over a large area. The traditional method has always been to organise rounds by meter inspectors – an extremely expensive system. Visits cannot be made very frequently or carried out in anything like real time, which would be the operator's ideal solution from a financial point of view. One last disadvantage is that data recorded by meter inspectors have to be transferred into the central databank. All the networks are relying very much on new information technologies to make some headway towards solving this awkward problem. One suitable model might be the telephone invoicing method, with consumption automatically recorded at central switching level.

Experiments into remote-controlled meter-reading (usually through the telephone network) have been carried out on electricity and drinking-water supply systems. For the moment, they appear to have encountered organisational problems. Despite this, more experiments have immediately been launched. In Nancy and Rouen (France), for example, operators have equipped their staff with portable terminals. Not only does this mean that details of the meter-reading round, programmed in advance by microcomputer, can be loaded – in some cases by remote control – into the portable terminal but also that invoicing data recorded on the terminal at each meter can be downloaded directly into the microcomputer.

A similar problem has been raised in videocommunications cable network management. The user, who wants freedom to choose and a wide variety of programmes, is reluctant to pay an all-in price for cable television, or even for a set of channels. It would be quite easy to widen the choice of programmes, providing that a differentiated pricing system could be devised. This entails knowing who is watching what (pay per view), which in turn presupposes a minimum degree of interactivity on the cable networks. A few years ago, this was thought feasible only if fibre optics were used. Today, however, co-axial networks can offer some interactivity. The United States is currently installing the system on conventional networks. It will be easy for countries that have chosen fibre optics cable to do likewise. But by far the most interesting example is Canada's VIDEOWAY. Supplied through the cable network, VIDEOWAY uses a simple box that enables the subscriber to interact with television programmes. For instance, he can select the camera position for a live hockey match, choose a happy (or sad) ending to the TV film he is watching, and of course play interactive games, *etc.* On the one hand so user-friendly, the system also enables the operator to invoice viewers on a detailed, pay-per-view basis thanks to real-time viewmonitoring.

So a very wide range of new information technology applications are appearing in the field of rational network management. While examples and applications abound, many of them in operation for several years now, the lingering impression is of a technology still in its infancy, or rather suffering from teething trouble.

Hall's remarks recalled in the first section are particularly relevant. Network management still does not function as a real information system. However, although urban network computerisation is in the early stages, applications that are no longer confined to a single network but attempt to build "bridges" across to information in other sectors or other forms of logic suggest that this is on the way.

Cross-sector applications

Two major types of application are currently being developed in the cities furthest ahead in this field. Firstly, there are the information transfer systems that cover several networks simultaneously and, secondly, geographical information systems (GIS) used to locate elements in several different networks.

The first type of application is based on the principle that in order to run networks – water, sewage, urban heating or car traffic – information must be obtained at different intervals from various parts of the city (sensor measurements, video images, voice signals, miscellaneous data) and then sent to central control units. Conversely, remote control instructions must also be sent out to different points in the city. At the moment, this information is generally sent down dedicated telecommunication lines belonging to individual networks, or through the public communications system (switched telephone network). But the technical requirements (such as reliability and flow capacity) for these transfers, together with economic factors (the high cost of dedicated lines used irregularly) mean that existing solutions are not always satisfactory. It was therefore suggested that a dedicated network be created for all this information. A "network of networks" would ensure reliability. It would be a "smart" system that could group data together, thereby saving on transmission and substantially reducing user costs. Furthermore, it could be tempting to take advantage of these networks and group together certain computing, memorising or monitoring operations carried out by individual networks. Without going as far as having a single centralised control unit, some grouping does seem possible.

The main issue is who would operate the network. Telecommunication regulations in some countries (France and Germany) give a dominant role to common carriers. In France, for instance, two projects of this type, one in Rennes and the other in Paris, have run into negotiating problems with France Télécom, which currently enjoys a monopoly in the telecommunications field and is in fact proposing its own network known as ASTARTE. Widely used for remote-controlled management applications (*e.g.* public building heating systems) or remote alarm systems, ASTARTE is extremely reliable. Furthermore, it offers adequate data-flow rates. Its prices are also advantageous for discontinuous data transmission.

In Germany, the Deutsche Bundespost has installed a similar network called TEMEX. The Münster network is used both for security applications (see below) and for recording and invoicing electricity/gas/urban heating/water consumption and for controlling the application of various electricity price rates (day/night, summer/winter, *etc.*). In Germany, several networks are commonly run by a single municipal enterprise (Stadtwerke), which facilitates applications of this kind. The same is true of systems that share the same infrastructure (service tunnel/ trench *etc.*). When a data transmission system is needed to run several different networks, it is easier to build it into existing infrastructure.

A European project (Italy, Spain and France with the support of the EUREKA Programme) for industrial infrastructure to carry the whole range of urban technical networks has been designed in such a way that a central unit diagnoses failures and other incidents at a distance and monitors the networks, using fibre optics built into the infrastructure.

Another interesting case is that of geographical systems for information on networks. This involves building and running a geographical database containing all the necessary information on components in the various networks serving a city. The systems are primarily aimed at facilitating any work required to build, maintain, repair or renovate urban networks, mostly on public thoroughfares. This means that they have a number of specific features:

- They operate geographically. The data system must be able to *locate* where any part of a network (pipes, joins, manholes, *etc.*) stands in relation to its environment, to (public/private) property and to other networks. A large quantity of data therefore has to be processed[6], and the need for geographic, cartographic and even videographic access to this information is obvious. This implies suitable information systems which, until recently, were extremely expensive. Today, data-processing systems (''object-oriented'' systems, workstations) give higher performance and now cost only a few hundred thousand French Francs.
- Services and operators are co-ordinated. The main advantage of these systems is that they are permanently updated, whatever the changes introduced in any of the networks. In practical terms, the data-processing applications for planning and following up work on the various systems must be linked up to the geographical data system. Thus any alteration to a network will immediately appear in the database, with no time-lag or additional processing. The same goes for any construction work that may affect public highways and any cadastral or urban-planning modifications, *etc.*

The fact that various applications are linked up to the general database in this way, presupposes that all those involved work in collaboration. In countries where the city authorities run most of the networks, they will naturally take on responsibility for co-ordination. Elsewhere, it may be more difficult as there is no single operator. However, the fact that an operator or supplier belongs to the public or the private sector should clearly not be a barrier to co-ordination, as Denmark's report for the OECD rightly emphasized.

Despite the problem of co-ordination, these applications are so invaluable to network operators, planning departments and city authorities in general that they have progressed very rapidly. In reports to the OECD, most countries – Greece, Canada, Portugal, Norway, Denmark, New Zealand and Finland – mentioned the creation of similar systems. On the other hand Turkey made the point that its data systems were still based only on individual networks, *i.e.* that there was no link-up to other network databases, to underline its hope that it would soon be able to boast its own co-ordinated information systems.

It is therefore highly significant that cross-sector applications for new information technologies are becoming widespread. For a long time now, the need to co-ordinate urban networks has remained just a catchphrase for public authorities, with no action being taken. Today it seems that a move is definitely under way, for several basic reasons. Cross-sector applications demonstrate beyond question the extent of the progress made in information technology by giving outstanding value for money. But above all, as mentioned in the first section of this paper, it is precisely through the concept of information implicit in new technologies that what has always been empirical knowledge can now be turned into something tangible and transferable, so to speak. Finally, the rise in car traffic, together with heightened awareness of environmental damage, has made repair work on the road system and even more so on private property[7] a prime concern in city and urban network management. In fact this key problem forms a ''bottleneck'' in the urban management process, which cross-sector applications of new information technologies do go some way towards removing.

Security applications

In this field, as in others, many applications merely serve to improve existing services. For instance automatic signals may replace telephone calls; or a monitoring system might replace an observer without any new service actually being created. Police, fire and medical services have obviously benefited from improvements in telecommunications and data-processing as they already have done from the telephone. However, we can safely say that the change wrought by the telephone was much more radical than improvements currently being introduced. The telephone in fact already brought with it interactivity and virtually instant transmission, both of which are particularly necessary for urban security. Subsequent improvements have so far produced suitable computerisation for managing data (particularly statistics) and some automation. Nevertheless, these systems are particularly useful when there is even a remote degree of serious risk because they avoid mobilising special staff to prevent events that rarely occur. Flood prevention is a case in point. Measuring water levels (or flow rates) upstream in the various tributaries of a river flowing through a city is one way to prevent floods that would be particularly harmful to urbanised areas. Various countries long ago set up forecasting systems using people as observers who contact security headquarters by telephone (or in some cases radiotelephone). Advances in sensors, microprocessors, and data transmission/processing technology have led to extensive automation, making the systems very reliable at a very acceptable cost. The same applies to remote monitoring services for environmental applications. One example is satellite monitoring for the road transport of hazardous materials, a service described in the report submitted by Finland. In some cases, however, the highly specific nature of the information and security arrangements involved means that the lack of any human presence may raise a problem. So there are relatively few important applications in this field, and these fall into three categories, namely the extension of security networks (more points, or increasingly remote points, being connected, *etc.*), the use of video filming, and the development of "smart" networks.

Norway's report is a good illustration of the first two categories. The widespread use of centralised video monitoring in public places and building entrances as a crime-prevention measure shows one application of video technology, now much cheaper, and the systematic control achieved by extending a network. Finland is developing similar applications. In France, the city of Nice has created "INFO-ALARM" along the same lines.

Network intelligence is called upon when alarm signals have to be memorised, emergency calls switched through to the appropriate officials (according to availability), or automatic tests and checks carried out to prove that calls are genuine. Different parts of the network can be "intelligent". With some applications, the terminals memorise and carry out instructions; with others, the whole network is designed to fulfil all the functions, using a switch mechanism for instance. In the first instance, there are the devices, supplied to elderly invalids, that automatically place a call through the normal telephone network. In Finland, where municipalities are obliged by law to provide a "safety phone" network for those who need it (the elderly, the housebound or invalids), automatic calling/dialling systems, special microphones and loudspeakers have been developed by private firms and launched in every city on the initiative of the Ministry of Social Affairs and Health. The second category includes the French ASTARTE and the German TEMEX networks (see above).

The most significant security applications are no doubt those that link up user homes to information, communication and emergency services. Here, genuinely new services have been created along lines that differ from any conventional emergency services. The broad trends described in the first section of this paper explain how these applications came about, why they were successful and why new services were created. Because urban society is becoming fragmented and scattered, because neighbourly relations are fading and thereby creating a strong demand for direct linkage to networks, and because at the same time people are attaching greater value to security in modern society, the organisations traditionally responsible for these issues no longer measure up to the task. Without exceptionally generous resources, firemen, police, social workers, nurses and doctors can no longer respond to demand which takes many forms and is urgent and virtually permanent, requiring not only that people and their property be protected, but also that they be made to feel less isolated. Although demand for security concerns public places (open areas, streets) and services (public transport), in many cases it also focuses on the home, the ultimate place of refuge and protection. The first alarm services to respond to this demand naturally made use of the normal telephone network. However, prevention being as important as cure, this was found to have limited potential both from a user and a service point of view. So new information technologies have been called upon to extend available capacity. Subscriber homes have been fitted with devices that transmit automatic signals (burglary, intruders, accidents, emergency signals in case of illness or inability to call for help, etc.). To quote but one example, electrocardiograms can now be sent through the system when heart patients need to consult their doctor urgently. Networks are designed so that any information sent through them will be properly directed, analysed and summarised, and an initial diagnosis can be carried out for appropriate and effective intervention. It is very important, for instance, to be able to locate the call correctly.

The networks in use fall into three categories. The most popular is the switched telephone network, which now gives better performance and can easily be connected to automatic call/reply devices, or used for different types of transmission (voice, data and, soon in some countries, fixed images). One French application, for instance, enables the elderly or invalids to call for help if feeling unwell, simply by pressing a button on a pendant they always wear around the house. The device transmits a radio signal to the person's telephone. With the help of a memory and a microprocessor, the telephone dials a series of preprogrammed numbers until the call is answered. A prerecorded message is then played and the call located. In Canada, the province of Ontario has developed a similar system.

Another trend is to allow for the fact that the switched telephone network may be inadequate or unsuitable, particularly when it is vital to make immediate contact without permanently ''monopolising'' a line (dedicated lines, for instance, are considered too unreliable and expensive for ''consumer'' applications). So use is made of existing telephone infrastructure, to carry a new network with different functions and rates. This is the case in Germany with the TEMEX network, used by the city of Münster to operate its Altennotruf-Münster service for the elderly and infirm. Here too the user is given a portable device and put through to an exchange, then connected to the person or service able to provide the most rapid assistance (doctor, neighbour, etc.). In France, the town of Douai uses the ASTARTE network for its security service, offering optimum response from public services in case of emergency.

Finally, a third and more recent trend is towards using infrastructure other than the telephone system, namely video cable networks. While these can clearly be used on a ''top down'' basis, warning their users of risks they may run, two problems are raised, namely how to extend the networks and how to ensure ''bottom up'' contact, without which subscribers do not feel completely safe. Frequently-quoted examples are Tama and Igashi-Ikoma, Japan's experimental ''wired cities''. The first is equipped with a co-axial network, the second with fibre optics. In Tama, 500 households were linked up to the network and in Igashi-Ikoma 160. There is no question of extending the service since both are experimental. While Tama's co-axial network was mainly designed to provide services on a ''top down'' basis, in Igashi-Ikoma services are interactive. The principle behind security services has been to establish a local emergency information channel designed to warn the inhabitants of any threat of oncoming disaster, such as typhoons, earthquakes or pollution. Information is given on where the incident is likely to occur. Nevertheless, this would not be very ''reassuring'' (quite the opposite in fact) if the system did not offer its users a chance to call in for information, forecasts, advice and help. The Tama network provided the public with only limited scope for interaction (a choice of television programmes) and subscribers were obliged to turn to the telephone network. On the other hand, the Igashi-Ikoma system, called HI-OVIS, includes from a single terminal every possible form of communication with the outside world (voice, data and even video imagery). There is no doubt that it offers real potential for aid/emergency services, while simultaneously creating a strong sense of security thanks to the type of link it offers with the outside world when problems arise.

Here too, the general impression given by these security applications for new information technologies and examples of their use is one of a technology that is still in its infancy. Needs appear to be substantial[8]. The services on offer so far have been a resounding success.

However, before these networks can spread throughout every city, like the conventional networks installed during the last century, a certain degree of both technological and organisational maturity will have to be achieved. The examples described here will contribute substantially to this.

New information technologies – repercussions and future trends for urban network management

Current impact of new IT on urban network management

As new information technologies have by definition made a recent appearance, most judgments remain very guarded as to their impact on the management of urban or other networks. OECD surveys among Member countries have revealed that more hindsight is needed before an opinion can be expressed on the direct or indirect effects of a given innovation. Certain conclusions can however be drawn – albeit temporarily – from three different standpoints.

Firstly, from an economic point of view, a tentative cost/benefit analysis of these experiments can be made. On the cost side, taking only those applications strictly confined to network management and the costs relating to new equipment (thereby excluding for instance the total cost of a fibre optic network dedicated primarily to television broadcasting, or the cost of acquiring GIS data that has already been collected), the figures are remarkably low. For the operator, they rarely exceed a few hundred thousand Francs. These figures should be set against the usual investment levels for technical networks (civil engineering, pipelines, equipment, *etc.*). In the rare cases when a comparison can be made of the investment costs incurred to extend a network and new IT investment costs with the same aim, *i.e.* optimising management, the latter account for only a few per cent (less than 10 per cent) of equivalent civil engineering investment. What is more, with increased standardisation, enhanced performance, better equipment (computers, transmission lines, *etc.*), these costs will actually decrease[9]. Consequently, the actual cost of an innovation will rarely act as a brake on its introduction, whether for experimental or for production purposes[10].

As for the benefits of new IT, the issue is much less clear-cut. Where operating costs are concerned, information technologies aimed at rationalising network management and matching supply to demand can lead to substantial savings on products (water) and energy (urban heating).

However, many applications are aimed, implicitly or explicitly, at increasing productivity. This of course is very hard to measure. Even so, it would seem that any savings there may be are not easily identifiable. This is because savings on staff, in the short term at least, rarely take the form of a fall in salary costs. Even if this were the case, it would be very difficult to transfer the savings into other budget item resources, given the principles on which most public networks are managed.

While new services (*e.g.* security services) do appear to offer operators significant long-term benefits, current operating conditions (innovative technologies, few subscribers, service constraints) mean that they are still not very lucrative.

Basically, it is as if new IT applications for urban network management were developing in a world where strict economic laws had momentarily been relaxed: costs are low (and hard to assess), benefits are low or inexistent (and equally difficult to assess). Of course the situation will not last. The next stage is expected to be the introduction of technologies that are more use-specific, more integrated and most probably more expensive despite falling component costs, and the widespread development of services, involving high investment but also particularly lucrative economies of scale and network effects. Much the same situation was faced, for instance, when it was decided last century to introduce arc lighting in main city squares. The firms that launched the innovation encountered financial setbacks but, several decades later, incandescent then luminescent electric street-lighting was introduced everywhere and became a resounding financial success for the operators.

So the idea is to cross thresholds and develop synergies between networks and achieve avalanche effects that will swiftly bring dramatic increases in productivity.

From an organisational point of view too, the impact is still low. New information technologies have the potential to re-organise existing services or create new ones. In practice, however, most of the applications reported are aimed at improving existing services but not at re-organising them or at creating new ones. In the meantime, apparently, new information technologies are not being used to "bulldoze" existing arrangements. This can be put down to some form of inertia or reluctance to change. Canada's report referred discreetly to wary middle-management and wavering senior-management attitudes. But it is particularly interesting to draw a parallel with the situation in industry. In large Western manufacturing firms, new information technologies can be said to have provided a great opportunity to undertake radical change. Generally speaking, continuous production, just-in-time and zero stock management methods all owe their appearance to information networks and data-processing systems. They have brought about drastic upheaval in work organisation. But these industries then found that they were facing crucial challenges: global competition from countries with very low labour costs, heavy competition from countries like Japan where work is organised along entirely different lines, and the need to respond to highly diversified markets. While urban network management also faces challenges, they are of a different nature. Competition is the exception, and new trends in demand take longer to surface. So new IT applications are not yet viewed as a strategic opportunity to introduce new forms of organisation better suited to the present situation[11]. Consequently, the organisational impact has been very limited and the process of reorganisation essential to some applications (automation of certain functions) attenuated and spread over a longer period, which explains why it is not easily observed.

So if there has been no great economic or organisational impact, do new information technologies have any other effects that might explain why applications are proliferating in most cities throughout OECD countries? Their main success has no doubt been to raise or at least maintain user service quality where there might otherwise have been a decline. The fact that networks are strongly identified as a symbol of equal treatment for all towndwellers (already mentioned early on in the paper) coupled with changing urban patterns and attachment to individualistic values, *etc.* could well bring on an urban crisis through the gradual decline of services provided for citizens, or services failing to live up to expectations (instant "real time" response, *etc.*). So far, however, this is not the case and the new IT applications described earlier appear to be going some way to postponing it by significantly improving the situation, or at least preventing it from worsening too rapidly. Examples abound. For electricity supply networks, peak periods pose a special problem. Pricing may be one solution, but it has its limits, such as the rules on price balancing to avoid widely differing user prices around the country. On the other hand, differentiated rates for consumers who are willing not to use power during peak periods can be effective, particularly if it is also technically possible to cut off supply at a distance on virtually a "real time" basis.

In fields where pricing is not or cannot be practised, or is socially and politically inacceptable, new information technologies are of great help to network management, enabling heavier flows to be carried without overinvestment or resorting to regulatory pricing. Examples include rainwater systems in many cities in the United States, Japan, France, Germany and the Netherlands, waste collection in German and French cities, and lastly all kinds of urban emergency systems (accidents, illness, disasters) which there is no question of regulating by economic means.

The public, elected representatives and operators have realised this and accepted these innovations, judging their impact to be beneficial. But here again, the situation will not last. Partial or one-off innovations will not go on solving urban problems indefinitely. The only answer is for genuinely integrated urban systems to become widespread. Hall, mentioned earlier, pointed out that it took some 60 years to move from the Model T Ford to an automobile-based system. Might not technologies mature faster so that current applications could develop into really integrated and widespread urban information systems suitable for networks? It cannot be ruled out, in view not of applications as they exist today but of the strong trends already indicating the path they will take in the future.

Future prospects

This section will be dealing, on a necessarily more subjective basis, with the way applications will no doubt be developing in the medium term.

Matching real and virtual networks

In most cases, it was the historical setting prevailing, particularly technical and economic circumstances, when cities first created their water, energy and other systems which actually determined that services would be provided on a "top down" basis, *i.e.* once the operating principles and production methods had been drawn up, a service was then carried through the infrastructure and provided equally for one and all. The only transaction involved was consumption invoicing.

In modern cities, however, another model is beginning to emerge. The sum of overall user demand plus overall supply, located at points throughout the city, gives a set of connections representing at any one time what might be called intended transactions and thereby forming a "virtual" network. This is the "demand-responsive service" so familiar to urban transport specialists, but the concept can be transposed to other networks. One feature of virtual networks is their high connectivity, *i.e.* a large number of point-to-point links can be envisaged that may or not actually occur, depending on whether the transaction takes place or not. The telephone system is in some ways the archetypal virtual system.

For obvious technical and economic reasons, existing networks – given their infrastructure – do not match up with virtual networks. For the future, one very strong trend appears to be a move to match real and virtual networks more closely together, at least where service provision management is concerned, thanks to new information technologies. Being able to analyse demand in real or almost real time by finding out how many transactions are expected and where they will be located could mean a continuously updated service provision network, thereby matching supply to demand as closely as possible.

Some experiments have already been set up, apparently achieving success in a variety of fields ranging from water supply or public transport to solid waste. Examples in Münster (Germany) and Douai (France) show, as we have seen, how emergency call demand organised the communications network in such a way as to provide the most appropriate form of assistance. In the Paris area (France), the Compagnie Lyonnaise des Eaux, a water facility, uses a model to optimise connections between the sub-systems that it operates in separate districts or groups of districts. The model makes it possible to plan up to several hours in advance how water will be moved between reservoirs in the various systems, so that these can be filled sufficiently during the night (when water use is at its lowest and electricity for pumping at its cheapest); to do so, it takes account of demand forecasts, required stocks, available resources and pumping costs between reservoirs. So the whole network is reorganised daily according to a virtual network for water use in specific parts of each subnetwork.

Another interesting example, though in the field of public transport which is not under consideration here, is the CRISTOBALD system[12]. Developed by URBA 2000 in an urban area of south-west France, the system enables would-be passengers to call from a touchtone telephone keypad in their home and indicate to an exchange the journey they intend making by public transport. In fact, they are not entirely free to travel where they choose. There are routes but they are virtual ones that only exist if they are explicitly requested. To create a virtual route, for instance, a bus can change routes; another vehicle (*e.g.* minibus) or even a taxi may be sent if the number of passengers does not warrant a larger vehicle. The advantages for the network operator are obvious, since he can satisfy user demand at a reasonable cost, even though users live quite far apart. For users, the main advantage lies in the fact that the system is reliable and easy to access, thanks mainly to an electronic calling and routing system. Although in this respect it is still experimental, CRISTOBALD is without doubt a precursor of future network operations.

A similar example based on more closely matched virtual and real networks is to be found in Germany. In Frankfurt, a waste exchange operating on the Bildschirmtext network (the equivalent of the French Télétel system) permits a large number of transactions between suppliers, who describe what kind of waste they have available, where it is and so on, and waste recovery specialists who also give relevant details. Depending on their transactions, a collection network is formed according to the type, quantity, location and cost of the waste involved.

These examples go to show how new information technologies can be used to update urban networks over the very short term and match them more closely with virtual networks[13]. There is no doubt that the range of applications can be considerably widened; they could:

– cover urban areas which will count as a single unit for the purpose of service provision, whatever the distances and population density involved;

- spread to a large number of cities once pilot projects have been completed; and
- open up to include most existing services.

New information technology applications involving teleshopping are particularly relevant at this point. Until now, commercial distribution was not considered as a potential urban network. It seemed to require the physical presence of the purchaser, the sales staff and the products together at a point of sale (the shop). But as a result of rapidly evolving distribution techniques, of the distance that the public must cover to find available products and, above all, of advances in new information technologies (image transmission, information services, tele-transmission), change in this field is not to be ruled out as teleshopping continues to develop. Various examples have been reported in the United States, the United Kingdom, Canada, Norway and France. Techniques vary, as do the areas of use. Conventional television, cable television, videotex, fax, digital telephones or combinations of all these can be used. In the United States, for example, certain commercial distribution chains will prepare in their super-markets (for a service charge of $10) any order that a customer sends in by fax and picks up by car. Smart cards can be used to identify the purchaser, confirm the order, pay for the goods, *etc.* These include everyday consumer items, frozen foods, cooked meals, second-hand cars or services such as rail ticket reservations, fast food meal orders, *etc.* For teleshopping services to work, products must of course be distributed through a physical distribution network. But this is always designed[14] around the delivery points for the orders received by the information system.

There is vast scope for this kind of application. Public-sector operators are primarily concerned with setting up information systems that may conceivably not be specific to certain types of purchase, nor even to teleshopping, but extend to vast network management systems, in which it is vital for virtual and real networks to match up very closely.

Exploiting technical potential to the full: imagery in network management

With many current experiments, the lingering impression is that new information technol-ogies are not being exploited to the full. This can be put down to concern about reliability or economic constraints, but also perhaps to unfamiliarity with the most innovative aspects of new technologies. It is surprising to see from the OECD survey, for instance, that expert systems are rarely mentioned in conjunction with urban network management, and yet they can often bring about significant improvements in operating performance. These obstacles should doubtless be rapidly overcome in the future. Close attention must therefore be paid to what current technol-ogy, exploited to the full, can offer urban network management.

This being so, it would seem that the most promising area is image transmission and processing. This field is currently overshadowed by the substantial stakes involved in television and audiovisual techniques. But in the specific field of urban network applications, there is considerable room for development. Canada's report gave an excellent example of this. The city of Toronto has set up an information system whereby a central unit guides firemen to the site of a fire. Any time gained is particularly precious and guidance is a great help to firemen who are unfamiliar with access routes to and inside the building which may be hidden by smoke or blocked by traffic, crowds, *etc.* They are guided by using a three-dimensional graphic system projected onto a screen from an updated data-bank. Here, graphics are a definite advantage, despite particularly demanding conditions for use (the databank must be perfectly up to date, the operator must be highly competent in guiding firemen by radio, and some organisational change in the normal running of the service is inevitable).

In the future, similar applications using computerised imagery are very likely to be developed for network management. Examples in industry have shown this to be possible. The nuclear fuel reprocessing plant in La Hague, France, for instance, is now run using HAGUENET, a fibre optic network. This uses high-density flows of digital data to create, almost instantly, fixed 3D-images of those parts of the plant to which staff do not have access for safety reasons. On a more general level, transmitting fixed and even animated images will become very economical over the next few years, thanks to compression techniques currently undergoing successful tests. Image transmission systems of this type will become very valuable in fields ranging from remote maintenance and monitoring systems to security applications, remote alarms, surveillance and teleshopping, but it will raise the problem of how to use the images produced. The high-speed visualisation and use of images cast up on a whole series of screens soon becomes too much for the staff employed to monitor them. Automatic systems must therefore be developed to process, summarise and compare image-based information, draw attention to anything unusual and trigger action or alarms. Current progress in the field of shape recognition lends added credibility to such developments. As an illustration of this, a large French motorway company is already planning to replace its usual real-time induction-loop system for measuring traffic with banks of video cameras that feed images into a central unit. Using the images, a shape-recognition system can count the different types of vehicle, measure speed and monitor for problems. If necessary, of course, the images can always be directly monitored by a member of staff.

Imagery also has potential applications for network users. The Japanese experiments appear to show how valuable this form of communication can be for security management. But image transmission networks might conceivably be developed for the remote maintenance of domestic appliances, which would thus be the network ''terminals'' (boilers, taps, *etc.*). As a last example, images could be displayed in public places to indicate the state of operation of urban networks (information about road traffic conditions, public transport availability in the rush-hour, rainwater drainage, *etc.*) and would be a way of encouraging the public to become more involved in equipment and flow management, all of which is an essential part of city life.

Finally, a third trend appears to be the development of real time management based on new information technologies. The advantage of this method for a number of network management problems has been demonstrated throughout this paper. Apart from organisational difficulties, the barriers encountered so far have been difficult and costly high-density transmission, together with excessive time lags in information processing. These barriers are likely to be overcome during the next few years, however, as transmission networks designed technically and economically for high-density flows became widespread and as processing lags are reduced. Furthermore, network management tools with current time lags ranging from a few minutes (*e.g.* about ten minutes for rainwater drainage) to several hours (for water, urban heating, electricity) might well eventually become real-time management software programmes. This will have a considerable impact in terms of both cost-effectiveness and service quality. Consequently, despite the organisational changes that would ensue, operators are very interested in any advances in this field.

The three trends presented above, namely more closely-matched virtual and real networks, maximum use of imagery in network management and development of real time management, could be hindered both by economic considerations and by organisational issues. They will not be considered as more than just a handful of innovative experiments unless network operators find them economically and strategically advantageous. For the moment, this is not the case for conventional networks (water, sewage, energy, *etc.*) or new networks (security, commercial distribution), at least as far as the operators are concerned. None of them has sufficient mastery of the information environment used in potential applications to launch a full-scale offensive on this front.

Nevertheless, there is virtually no doubt that the information environment itself will change over the coming years. The point here is that many countries are planning to extend their integrated information and communication networks. Integrated Services and Data Networks (ISDN) are leaving the drawing-board and entering the real world. This is due to action on the part of telecom network operators who see ISDN as the successor to telephone systems, now that a wide range of services require digital data transmission. Deutsche Bundespost Telecom in Germany and France Télécom in France are now well committed to the process, whilst US and UK operators are carrying out tests and the Japanese are planning to set up the networks in urban areas by the year 2000. Basically, ISDN designers have acknowledged that the data to be sent will be in digital form[15] and have created a network that is efficient enough right down the line, *i.e.* right through to the subscriber's terminal, to transmit any data (including images); it will also ensure sufficient interactivity and multiple network configurations for each service provided. Unsolved problems still remain as to the choice of carrying medium and the choice of terminal devices to integrate/disintegrate the information. As far as the medium is concerned, monomode optical fibre appears to be gaining ground and distribution to homes in urban areas may cost as little as telephone distribution[16]. The price of terminal equipment, vital to integrated systems, is also expected to fall with the advent of mass production using standard components. The European Community's RACE programme is offering to carry out co-ordinated research, experiments and standardisation in this field.

For ISDN operators, research in various countries has shown that considerable economies of scale can be achieved, while ensuring performance that will satisfy a wide range of users. In particular, it is planned to offer subscribers the possibility of sending out their own information in any form (including moving pictures) except high-definition television, which would be reception-only. Management applications for urban technical networks will therefore be free to transmit anything from digital information to fixed or animated images with a quality similar to that of present-day television broadcasting. Finally, the smart card could be a valuable addition to these systems. It will facilitate any operation involving subscriber recognition between terminals, confirmation of transactions or the downloading of basic input from mobile (transport) or portable terminals, *etc.*

The strategy adopted in setting up ISDN varies from country to country. With digital switching equipment at their disposal, Germany and France have lost no time in connecting a large number of subscribers throughout each country. In Germany, 135 cities will have been connected by 1990, with access available throughout the country by 1993. France will already have achieved this by the end of 1990. Applications have already proliferated: data exchange is used in France for planning permission and in Germany for transmitting medical imagery. Other countries have begun by installing fibre optics along the major routes in their telephone network and then extending ''fibre'' (and therefore integrated services) right through to distribution points (United States). Finally, other countries with high cable penetration are planning to use cable broadcasting networks to carry ISDN. Although it takes a number of forms, the move appears to be under way and irreversible. The expected outcome is for almost every workplace and home, at least in urban areas, to have access to ISDN. If so, the network management trends described above may make their impact felt much sooner. This would be a step towards an integrated urban information system that would become, for citizens and urban authorities alike, a basic feature of city life and urban management, just as housing and cars are today.

RECOMMENDATIONS

For the moment, as we have seen, it is difficult to take stock of the use of new information technologies in urban network management. Applications are numerous and promising but still only case-specific. It is not an easy matter to draw up the broad criteria needed to carry out a cost/benefit analysis that would go beyond individual cases.

On the other hand, future prospects are of the utmost interest. We cannot rule out a radical move by technical networks towards "customised" services, following the development of new information technologies. Finally, network management potential has still not been fully exploited and may lead to major developments in the near future.

What recommendations can be addressed then to those in charge of urban management, for whom urban policy issues are more important than technological specifications? One point to be stressed is the degree to which society accepts the use of new information technologies in the management of urban networks. This is what will determine whether or not technologies will be used elsewhere than in the individual experiments and sector-specific applications considered in this paper.

There are two quite distinct aspects to social acceptability. On the one hand, new information technologies must be acceptable to those in charge of operating the networks – workers, office-workers, technicians, engineers and executives responsible for the creation or day-to-day management of water/electricity/sewage networks, systems, *etc.* No doubt this will only come about provided that there are obvious "system effects" to prove how advantageous new information technologies are. Such signs begin to appear when there is synergy between information or communication systems in several different networks. The telephone is a case in point. Geographical information systems are beginning to follow suit. Videotex in the United Kingdom, Canada, Japan and Germany and particularly in France is also expected to play this role. The advent of ISDN should make the impact even more noticeable. But none of this will suffice if new information technologies are not given systematic consideration by all those who take part in network management, *i.e.* public services, technical departments, public and private enterprise. The only way to ensure this, however, is to draw up and accept common objectives that will only be achieved if new information technologies are used. Just as certain urban authorities have reorganised their departments by computerising them on a "paperless management" basis, so a "zero defect" target for all municipal network services would promote the widespread use of new information technologies, which in turn would eventually result in substantial "system effects".

Social acceptability also applies to users, in this case towndwellers. So far, it is clear that, for the networks relevant to this study, they are not particularly interested in new technology for the sake of it. In most cases, they do not realise how new information technologies are used in water supply or household refuse collection. And yet it seems clearly very important to receive their backing for this type of use. It is the only solution if certain thresholds in new IT development are to be crossed. Long-term support will not be obtained solely through the public liking a device because it is modern or an experiment because it is spectacular. The point to bear in mind is that new information technologies should always be introduced to network management with the aim of giving real power to the user as a towndweller (or the consumer as a customer). One aspect of this is the degree of power (however small) to command and control that the user as a towndweller can exercise over the network. Another aspect is how reliable such command and control will be. As long as new information technologies appear as the condition on which a certain amount of power can be given back to the user as a towndweller, if they can help him escape the impersonal, mass-produced, standardised system of network supply/collection/distribution, if they bring hope of more personal treatment, better information, real freedom of choice, they are bound to be accepted. Surely urban networks are offering a new version of the kind of public participation often demanded in the field of town planning?

In any case, that will be the price to pay for a genuine system of urban "utilities" management based on the concept of information. To go back to Hall's metaphor, it is the requirement to be met if we are to move rapidly forward from the Model T Ford era to the modern automobile-based system.

NOTES

1. Nevertheless, operators still have to harmonize operating standards for networks that were originally in rural areas but now belong to urban areas. They will also have to plan for local (not just regional or national) quality control for their services, as provided in urban areas.

2. Urban development and impacts of technological, economic and socio-demographic changes. Report of an expert meeting, Paris, OECD, June 1987.

3. One-third of all projects carried out in the various fields of activity, ranging from health to training and leisure, *etc.*

4. It should be noted that the survey referred to "municipal" services only. However, many new ICT applications are aimed at managing city networks run by operators other than the urban authorities or at a higher level. "Electronic mail" is one example. If this kind of application were taken into consideration, the share held by the transport sector would balance with those of other networks.

5. This is why, as we shall see, "smart" cards are not widely used, if at all, in these fields, whereas they have become very popular throughout the transport sector and other services where the user plays an inherent role in service provision.

6. In Toulouse (France), the geographical information system is even linked up to the SPOT satellite image databank.

7. Canada's report mentions a databank in Calgary containing details of all pipes running under private property (*e.g.* under lawns).

8. There is indeed a very broad spectrum of needs. Applications not mentioned so far range from identifying telephone calls to the home (to avoid unwanted calls) to establishing electronic prisons, an idea suggested in the United States to solve prison problems. This would actually mean releasing "prisoners" who would be given electronic tags connected to a central monitoring unit that could locate them.

9. That is, if mass-produced components are used that are not 100 per cent reliable. Totally reliable components, *i.e.* ones that have been tested and guaranteed (military applications, automation in the aeronautics and rail sectors) are still very expensive.

10. This does not, as the Canadian report pointed out, prevent some local government departments from judging expenditure too high when resources are very limited.

11. With the exception, perhaps, of security networks.

12. Similar to the RUFBUS system in Hanover (Germany).

13. It should be noted that the problem is apparently the same, *mutatis mutandis,* in other fields less suited to the concept of networks. See bibliography to the article by Neil Wrigley on current developments in health care services in the Bristol (United Kingdom) area.

14. This is done in a variety of ways which at present may or may not use information technologies. In the future, however, this will be an extremely important factor when developing similar services.

15. Television picture transmission remains one notable exception where digitalisation is concerned.

16. Around FF 5 000 per subscriber, from the home to the first switching point.

BIBLIOGRAPHY

ANDERSON, D.D. (1981), *Regulation Politics and Electric Utilities, a Case Study in Political Economy,* Auburn House Publishing Company, Boston.

DROUET, D. (1987), *L'industrie de l'eau dans le monde,* Presses de l'Ecole nationale des Ponts et chaussées, Paris.

DUPUY, G. *et al.* (1989), *Génie urbain, territoires et informations, les annales de la recherche urbaine,* N° 44-45, December.

FEDERATION OF CANADIAN MUNICIPALITIES AND THE CANADIAN OCCUPATIONAL PROJECTION SYSTEM (n.d.), prepared by the Coopers & Lybrand Consulting Group, *The Challenge of Technology Change in Canadian Municipalities.*

FREROT, A., LATERRASSE, J. and ROWE, F. (1988), "Télétransmissions et gestion de l'espace urbain: quels opérateurs pour quels réseaux?" 10èmes journées internationales de l'IDATE, Actes, Montpellier, November.

HEPWORTH, M.E. (1990), *Project on Urban Impacts of Technological and Socio-demographic Change: Applications of Information Technology in Transport,* OECD, April.

LATERRASSE, J. and CHATZIS, K. (1989), "Evolution des réseaux et nouvelles technologies de l'information" in "Génie urbain, acteurs, technologies et territoires", Plan urbain, July, Paris.

LATERRASSE, J., CHATZIS, K. and COUTARD, O. (1990), *Information et gestion dynamique ou quand les réseaux deviennent intelligents,* Flux N° 2, Autumn.

OECD (1987), *Urban Development and Impacts of Technological, Economic and Socio-demographic Changes,* Report of an Expert Meeting, June, Paris.

OECD-URBA 2000 (1989a), *Applications of Information and Communication Technologies to the Fonctioning of Cities,* Paris.

OECD (1989b), *Auswirkunger des Technologishen and Sozio Demographischen Wandels auf die Stadtentwicklung,* Bundes Republik Deutschland, Nationaler Bericht, August.

OECD (1989c), *Project on Urban Impacts of Technological and Socio-demographic Change,* National Contribution from Finland, September.

OECD (1989d), *Project on Urban Impacts of Technological and Socio-demographic Change,* National Contribution from New Zealand.

OECD (1989e), *Project on Urban Impacts of Technological and Socio-demographic Change,* National Contribution from Turkey, December.

OECD (1990a), *Project on Urban Impacts of Technological and Socio-demographic Change,* National Contribution from Canada, February.

OECD (1990b), *Project on Urban Impacts of Technological and Socio-demographic Change,* National Contribution from Norway, March.

SPARMANN, J.M. (1990), *Project on Application of Information and Communication Technologies to Traffic Management,* OECD, October.

TARR, J. and DUPUY, G. (1988), Eds. Technology and the Rise of the Networked City in Europe and America, Temple University Press.

UNIVERSITY OF WATERLOO (1984), *Technological Futures and Human Settlements,* October.

WRIGLEY, N. (1990), *Project on Urban Impacts of Technological and Socio-demographic Change, Information Technology and Health Care Services,* OECD, October.

TELEWORK: VISIONS, DEFINITIONS, REALITIES, BARRIERS

Lars QVORTRUP
Professor at Odense University
DENMARK

Lars Qvortrup studied literature, linguistics and media and obtained his Ph.D. degree at Aarhus University, Denmark in 1974. He holds the chair of telematics and human interaction at Odense University, Denmark.

He has published a number of books and several articles on organisational communication, rural communication, community teleservice centres, etc.

He is editor of the Newsletter on Community Teleservice Centres, and he is co-editor of the European Overview of IT in Human Services.

His present research concerns IT and organisational interaction, technology assessment, and IT for rural and urban communication.

IF CITIES DIDN'T EXIST...

"If cities didn't exist, we wouldn't have to invent them." This statement by the American writer John Naisbitt highlights the ultimate perspective of urban planning – the disappearance of the city.

Historically, cities have always been the centres of production and capitalism. The city centre was the market square, surrounded by merchants' buildings and storehouses, and buildings such as the city hall and the cathedral, representing secular and clerical power. It had easy access to public transport. A good example is Copenhagen whose Danish name (Kobenhavn) means "the merchants' harbour" or "the trade harbour".

Under industrial capitalism factories were established in the centres of cities, and residential quarters and tenements surrounded the production area. Some English industrial cities are good examples of this.

Gradually, however, heavy industry was moved from the cities, which became administrative centres, dominated by huge private and public office buildings. Just go to New York and you will see what I mean. Early twentieth-century office technology, with traditional organisational thinking and growing administrative duties, led to the establishment of large, hierarchical administration complexes represented by skyscrapers with doormen on the ground floor and bosses at the top. Today such centralised administrative facilities are no longer necessary. The "remote office", with dispersed local work centres, is technically perfectly feasible, face-to-face communication being necessary only for a small group of managers and decision-makers, or for individual administration specialists who offer special high-grade services through telecommunications networks to many different firms (*cf.* Huws, Korte and Robinson, 1990, pp. 22-23, and Nilles, Carlson, Gray and Hanneman, 1976). If cities didn't exist, we wouldn't have to invent them.

But cities do exist. First of all because they are almost unchangeable historical monuments. Individual buildings can be demolished or their interiors changed, for example with the advent of new occupants, but it is hard to alter the basic structure of a city, the overall plan, the streets, the harbour. Often the history of a city can be read as you read the annual rings of a tree; new urban developments representing new historical periods grow up around the old ones.

And cities continue to exist because private and public administration is still dominated by old hierarchies, centred in traditional office zones. Remote work has not developed as fast as was expected, and distance workers are a minority of office workers.

Gradually, however, the "new city" must be invented or it will invent itself. With little traditional trading based on the flow of commodities into and from the city, with no heavy industry, and with a decreasing amount of traditional, centralised, bureaucratic administration, what will be the functional basis, the infrastructure and the architecture of the "post-modern" city of the future?

In this paper I will take a closer look at one of the basic symbols of current and future change in the urban structure: telework. The growth of telework and telecommuting will influence the infrastructure of future cities and the whole of urban planning, including dealing with traffic congestion, the design of residential areas, and the siting of office zones, shopping centres, schools and cultural institutions. It will also modify family life, reintroducing the household as a workplace and thus changing the roles of men, women and children.

After presenting some visions of the future, and discussing the problem of how to define telework, I will turn to the current situation. What is surprising is not the existence of telework, but its scarcity. Why hasn't telework been more successful in recent times? Answering this question will lead me into the final section: the identification of psychological and organisational barriers to telework. In the conclusions some suggestions for urban planning action will be presented.

Visions

The history of telework

The potential for telework is vast. Technically, all tasks which are unrelated to physical production can be carried out in satellite offices or district centres, or electronically in the home, instead of in central office buildings (*cf.* Steinle, 1988, p. 8). In 1987, an Empirica survey showed that a third of decision-makers in European companies were interested in telework options, and that 25 to 30 per cent of workers using IT equipment were interested in telework (Steinle, 1988; Huws, Korte and Robinson, 1990).

One of the first people to foresee the coming of telework was Norbert Wiener in his book *The Human Use of Human Beings: Cybernetics and Society* published in 1950. Wiener discussed a hypothetical example of an architect living in Europe but supervising the construction of a building in the United States. Using the recently introduced facsimile transmission service "Ultrafax", he said, "...even without transmitting or receiving any material commodities, the architect may take an active part in the construction of the building... In short, the bodily transmission of the architect and his documents may be replaced very effectively by the message-transmission of communications which do not entail the moving of a particle of matter from one end of the line to the other'' (Wiener, 1950, pp. 104-105). In conclusion, Wiener distinguished between two types of communication: communication as material transport and communication as transport of information alone.

Interest in telework did not spread widely until the early 1970s. The first reason was that the price of computers fell, making home computing a realistic possibility. Secondly, it became gradually understood that telecommunications and data processing would mesh into an integrated system. In the late 1970s the French neologism télématique (English: telematics) was coined by Simon Nora and Alain Minc in order to describe this growing interconnection of *tele*communication (télé-communication) and computers (infor*matique*) into one, integrated national and international system (Nora and Minc, 1978). Thirdly, the early 1970s was the period of the first international oil crisis, which focused critical attention on the waste of energy in private and public transport systems. The combination of technical potential and social need seems to be a precondition for technological breakthrough. The dominant view of telework in this initial phase was the potential to save energy by substituting electronic communication for physical transportation. According to Margrethe H. Olson the term ''dominetics'' was coined in 1969 in an article published in The *Washington Post*. In 1971, Frank Schiff began talking about ''flexiplace'', and around 1974 Jack Nilles introduced the concept of ''telecommuting'', the US equivalent of ''telework'' (Olson in National Research Council, 1985, p. 1). The first systematic analysis of the pros and cons of telework at a societal level was published in 1976 by Nilles and co-editors Carlson, Gray and Hanneman (*cf.* Nilles, Carlson, Gray and Hanneman, 1976). The human and economic costs and benefits of the employee travelling to and from work were compared with the costs and benefits of staying at home and using a telecommunications link to communicate with his employer.

Then in 1980, Alvin Toffler incorporated the idea of telework into his three-phase world history, making it one of the basic elements of the emerging ''Third Wave''. In his book *The Third Wave,* he predicted that the new information-based production system would move millions of workplaces from factories and offices back to where the workers had come from: the home (Toffler, 1980).

Social impact studies

In the same period the first sociological analyses of the impact of information technology were produced. A number of studies were carried out at the societal, organisational and individual levels. Today it seems that, even though many trends were correctly identified, the speed of development was overestimated.

At the societal level, as an example, in 1977 Bruno Lefèvre predicted ''...that the organisation of urban and regional space will be substantially modified, not to say turned completely upside-down'' (Lefèvre, 1977; his paper was later published in Forester ed., 1980). At the same time, Daniel Bell in more detail and in more realistic terms analysed the impact of the integration of computers and tele-communications on our concept of the city. In geographical terms, the location of high-skill workplaces, research laboratories, *etc.* is less dependent on the traditional factors of economic geography and more influenced by physical or electronic access to information and, in addition, the nearness of educational and recreational facilities. Visions had already been published of ''network towns'' or ''linear cities''. But, as I emphasized in the introduction, cities are inflexible, taking a very long time to change. However, according to

Bell, our concept of "urbanity" is being modified and "... the entire nation (if not large parts of the world) is becoming urbanised in the psychological sense, though increasingly more dispersed geographically" (Bell, 1979, p. 201).

At the *organisational level,* many studies have forecast radical changes to existing hierarchical institutions. Not only will computers in private homes affect the national and urban infrastructure; they will also modify the structure of private and public organisations. In his 1979 paper Daniel Bell refers to the change of libraries: "The image of the Alexandrian Library – the single building like the Bibliothèque Nationale, the British Museum, or the Library of Congress – where all the world's recorded knowledge is housed in one building, may become a sad monument of the printed past" (Bell, 1979, p. 191).

This idea was elaborated by Wilfrid Lancaster of the University of Illinois. In his book *Libraries and Librarians in an Age of Electronics* (Lancaster, 1982) he presented us with a radical analysis, forecasting not only "the disembodiment of the library" but also "the disappearance of the book". According to Lancaster, in the year 2000 the library will have been replaced by a computer network and the books will have been fully or at least partly replaced by so-called "machine-readable resources".

Still at the organisational level, a number of writers have speculated about the impact of telework on the working community. In the early 1980s, Starr Roxanne Hiltz put forward the concept of the "online community" which would gradually replace the physical office. According to Hiltz these office networks can best be thought of as a new kind of social system: one in which the familiar social processes of the workplace and the organisation are replaced by electronic online communities (Hiltz, S.R., 1981; Hiltz, S.R., 1984, p. 30).

Finally, at the individual level, a large number of studies have been carried out. While Toffler predicted a new, family-centred society, combining work and leisure activities, and breaking down old bureaucracies (Toffler, 1980), Neil Postman foresaw "the disappearance of childhood" as a protected and innocent phase of individual development, and a change of basic knowledge structures as a direct consequence of new media and communication technologies.

In the field of telework, J. Glover had in 1974 analysed the psychological, economic, and architectural problems of working at home, and at the same time P. Goldmark foresaw the possible establishment of alternative distance work organisations such as "community communication centres" combining computer-based work and business activities with teleservices from health and education departments, social services, and cultural and leisure institutions (referred to in the above-mentioned paper by Bruno Lefèvre, 1977).

Quantified predictions of telework

Some researchers and forecasters have even ventured into quantitative predictions about the future of telework.

In the 1970s and the 1980s very optimistic figures were published, predicting that within this century major sections of the workforce would be working from home (some of these forecasts are summarised in Korte, 1988, p. 374 and in Steinle, 1988, p. 7).

Still in the early 1980s, it was forecast that 10 or 15 million people would be teleworking two or three days each week by 1990 (Connolly, 1988, p. 3; and Cross and Raizman, 1986). At the same time the Institute of Future Studies at the University of South California was assuming teleworking by 20 per cent in 1990 and between 33 and 40 per cent in 2000 (Korte, 1988, p. 374; and Connolly, 1988, p. 3).

Even though expectations have been modified from year to year, in 1987 it was still a common belief among employed people in Europe that it would be entirely feasible to decentralise about two-thirds of all jobs, which would mean roughly 80 million jobs in Europe being carried out from the home or from locations near home (Steinle, 1988, p. 7).

After having evaluated the different proposed definitions of telework, I will return to the impact studies and to the quantitative predictions, confronting them with current realities.

Definitions

During the last 20 years telework has been defined in three different ways. These definitions correspond to trends in organisational structure, from the original centralised structure in which all administrative operations are located at a single site through three phases of increasing decentralisation [Nilles *et al.,* 1976 (Figure 2)]:

- Fragmentation. Relatively coherent sub-units of the central organisation break off and relocate closer to the homes of employees. Employees at these sub-units still belong to the same administrative unit in the organisation. The spatial evolution of organisations)
- Dispersion. A number of smaller work locations are established by the central organisation. Employees at a dispersed work location do not report to the department in which they work, but to the location nearest their homes.
- Diffusion. A number of small work locations or homeworking places are established in which one or more employees work for several different companies.

Fragmentation

Traditionally, telework has often been understood as "electronic homework", *i.e.* unskilled, low-paid office work from the home, based either on a full-time contract or on a freelance relationship with the employer.

Our image of this kind of telework is rather negative. Telework is associated with mothers, isolated from the working office community, doing monotonous word processing for a remote employer. Even if skilled work is added to this definition, it is still based on the assumption that there is a central office at which others are employed.

Originally, research into telework was influenced by traditional organisational thinking, and interest in it was based on the potential substitution of physical transportation by electronic transportation, *i.e.* communicating electronically instead of commuting physically. So in 1973 Jack Nilles coined the term "telecommuting" (Nilles, 1988, p. 301), and still today in American literature "telecommuting" is normally used for the European "teleworking". Thus, within this tradition, American "telecommuting" (in addition to Nilles, see Margrethe H. Olson in National Research Council, 1985, p. 1, and Cross and Raizman, 1986, p. 3f.) and European "telework" (*cf.* definitions in Maciejewski, 1987, p. 2f and Weijers and Weijers, 1988, p. 414) mean to work at a distance from a central office or headquarters.

I think that it is fair to summarise that, in this phase of research, the definition of telework (or telecommuting) is based on a limited fragmentation of traditional, centralised organisations. In addition, all definitions assume that computers and/or telecommunications are used. Some of the definitions identify telework with homework, while others include other places of work remote from the central employer. Finally, all definitions exclusively focus on the relationship between the place of work and the employer's centre.

Figure 2. **The spatial evolution of organisations**

(a) Centralisation

(b) Fragmentation

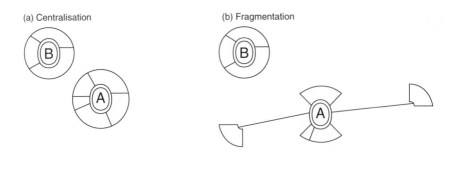

(c) Dispersion

(d) Diffusion

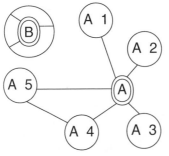

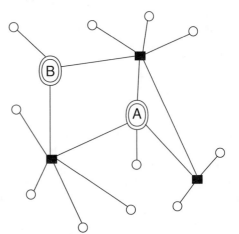

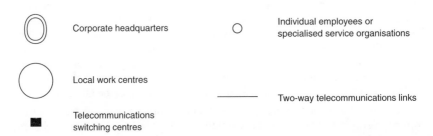

⬭ Corporate headquarters	○	Individual employees or specialised service organisations
○ Local work centres		
■ Telecommunications switching centres	———	Two-way telecommunications links

Source: Nilles *et al.* (1976, p. 12).

Dispersion

Though it may be debated whether telework is home-based or merely carried out at a different location, all the above-mentioned researchers agree that telework is done for a central office or employer. It is, however, not difficult to give examples of types of work which do not fit this definition, yet would be immediately thought of as telework. Some telework researchers, for example Ursula Huws (1988, p. 71), have realised this; she emphasizes that current, official definitions might exclude a large proportion of people actually teleworking from home.

For a start, many teleworkers are sole traders running their own, home-based businesses. They may be private consultants, working at home, sending reports to clients or to subscribers to a service they provide. They may be architects, working part-time at home, faxing or mailing their designs and plans, and commuting part-time in order to supervise the construction of a building in person. They may be information providers, selling to a mass market via, for example, a videotex system like the French Télétel.

Secondly, many teleworkers use telecommunications in order to avoid travelling to their customers or clients. Similarly, end-users subscribe to information services through the telecommunications network. An example is video-based training, and perhaps even home-shopping and home-banking belong to this category.

In these cases work is carried out remotely, and travelling – by the worker or the end-user – is replaced by telecommunications. Consequently, in addition to the above-mentioned substitution of telecommunications for the employee's journey to his or her central office, we must include substitution of telecommunications for the employee's or service-provider's transportation to the customer, and even, perhaps, for the customer's transportation to the service provider at home or at a central office.

In summary, this second-phase definition is based on the tendency of dispersion. In addition to traditional electronic homework, new businesses are established, using computers and telecommunications to serve their clients, and existing bureaucracies are building new "electronic" satellite offices.

Using this reasoning, a number of current researchers have concluded that it does not seem appropriate to define telework according to any single parameter or dimension (*cf.* Korte, 1988, p. 375; Steinle, 1988, p. 9; and Huws, Korte and Robinson, 1990, p. 9). Stern and Holti (1986) even argue that one is tempted to conclude that "the term has become so distorted and so lacking in conceptual meaning as to defy serious investigation". Consequently, instead of focusing strictly on "telework" one should consider "the diffusion and impact of these applications of NIT (new information technology) communications in the spatial organisation of work" (Stern and Holti, 1986, p. 7).

Diffusion

In the two above-mentioned phases, the definitions of telework have been based on the explicit or implicit assumption that somewhere there is a centre representing power, management, direction or just the physical office. However, in 1983 Turoff and Hiltz went one step further, arguing that current office technologies make the concepts of centralisation and decentralisation outmoded, and substituting a structure based on fluid networks by means of which teleworkers – or rather, they would say, knowledge workers – become members of *ad hoc* groupings formed around particular projects. These groupings – or "online communities", as they were later called by Hiltz – are not restricted by time or space, and are based on the specific knowledge projects in which they participate (Turoff and Hiltz, 1983; *cf.* Huws, Korte and Robinson, 1990, p. 32). Here, fragmentation and dispersion are supplemented by a radical diffusion of traditional organisational structures.

The best-known example of "networking" in "online communities" is provided by computerised conferencing systems used by groups of scientists working within the same research specialty. Typically such systems provide message systems which enable members to send private communications to individuals or groups on a topic of discussion, conferences which build up a permanent record of a topic of discussion, and notebooks in which text processing features may be used to work on jointly authored reports (Hiltz, 1984).

Based on such examples, Hiltz totally changes the definition of an office. "Usually," she writes, "one thinks of it as a place, with desks and telephones and typewriters. In thinking about the office of the future, one must instead think of it as a communications space, created by the merger of computers and telecommunications". According to Hiltz these computer-mediated communication networks can best be thought of as a new kind of social system, in which the familiar social processes in the workplace and the organisation become subtly altered by electronically mediated interactive processes, creating new kinds of "online communities" (Hiltz, 1984, p. XV and p. 30).

Recently, Huws, Korte and Robinson have accepted the same kind of definition. Looking into the future, they believe "...that the traditional concept of the workplace as a fixed geographical space will be replaced by more abstract notions of the working context as a set of relationships, a network, an intellectual space" (Huws, Korte and Robinson, 1990, p. 208). This "network office" is called The Elusive Office.

Summary

To sum up the search for an adequate definition of telework, one can specify three organisational phases (phases should not be understood literally, but rather as ideal types in the sense of Max Weber) and three corresponding definitions.

The first phase is characterised by an early fragmentation of traditional centralised organisations. The corresponding definition of telework is to work at home instead of in a central office, thus substituting telecommunication for the twice-daily commuter journey. The appropriate term for this is Telecommuting.

The second phase is characterised by a dispersion of traditional organisations. New, decentralised satellite offices and local work centres are established in addition to electronic homework. Still, however, the dominance of the centre is unchallenged. The appropriate term for this is Teleworking.

Finally, the third phase is characterised by a diffusion of traditional organisations. New specialised service organisations are being established, individuals work as advisers and consultants, traditional office work is replaced by work based on computer conferencing networks *etc.* The appropriate term for this is Networking.

In 1988, 15 years after launching the concept of electronically mediated distance working, Jack M. Nilles proposed a broad definition, covering all the above-mentioned forms of telework. He simply stated that telework "includes all work-related substitutions of telecommunications and related information technologies for travel" (Nilles, J. M., 1988, p. 301). Thus, the concept of telework is used for all kinds of electronically mediated work-related interactions across distance: interactions with the employer at the company headquarters, with colleagues in a central office or at other workplaces, and – as either an employee in a large company or an independent home-based or shared-facility-based self-employed person – with contractors, subscribers and customers.

In conclusion, in order to cover all the specific forms of telework described above, I think that the broad definition of telework presented by Nilles in 1988 is the most appropriate: telework is defined as all work-related substitutions of telecommunications and related information technologies for travel.

The organisation of telework

As already mentioned above, normally telework is associated with electronically mediated homework. However, telework can be organised in a number of different ways, and many research projects have suggested different classifications of organisational forms of telework.

The first proposed classification of telework in relation to different organisational arrangements was presented by Nilles *et al.* in 1976 (*cf.* Figure 2, above).

In 1988, Nilles refined his original scheme. He drew a distinction between two main forms of telework, or telecommuting as he prefers – firstly home-based telecommuting, and secondly regional centre telecommuting. The latter can be subdivided into three categories: satellite centres (centres set up by a large organisation to house only its own staff), local centres (facilities that house a number of teleworkers from different organisations), and neighbourhood centres (a mini version of local centres housing just a few workers, located within a short distance of the workers' homes).

Still in 1988, Korte put forward a new classification scheme, apparently based upon analyses and empirical results from a FAST research project between 1985 and 1986 on telework with participants from West Germany, the United Kingdom and France. Realising that Nilles' one-dimensional approach, based only on externalisation from the original centre, is not sufficient for classifying all current organisational distance work arrangements, he suggests using a two-dimensional classification approach, including not only location but also the contractual arrangement. An even more exhaustive classification by FAST research project's Stern and Holti in 1986 proposed six different categories of distance work arrangements. Like Korte's scheme, Stern and Holti's classification is based on two dimensions, including location as well as contractual arrangements. Moreover, in addition to the teleworker/employer relationship they also include the relationship between the teleworker and the client, thus conforming more closely to the broad definition of telework presented above. Almost the same classification is used in Huws *et al.* (1990).

If these classifications are combined, the following five categories can be summarised:

a) Electronic homework. In this form the worker undertakes paid employment either entirely at home or predominantly from a home base, with none or few visits to the site of an employer or client.

b) Shared-facility centres. In this form a building, office or workcentre is equipped with various new information technology facilities both for on-site work and for communicating at a distance. These facilities are shared by a number of users who may be employees of different companies, independent freelance professionals or small businesses unable to afford such facilities on their own. Shared-facility centres can be divided into:

  i) Neigbourhood work centres. Small offices or buildings equipped and financially supported by different employers and/or by public authorities. These centres are

85

placed in a residential area or in a single block and are used by no more than five to ten employees.

ii) Local work centres. Buildings equipped and financially supported by different employers and/or by public authorities. These centres are placed in a residential area and are used by a larger number of employees.

c) Satellite work centres. These centres are placed in a residential area, but they are owned by a specific company which has relocated part of its operations at a distance from an original or main site. The operations in the branch office are normally relatively integrated ones, and the branch personnel communicate with head office by means of new information technology.

d) Flexible work arrangements. Here, the distance workers may be located in more than one place: they are mobile, using portable equipment and telecommunication facilities, partly working at home and in trains and planes, partly at the central office.

e) Distance working enterprises. These enterprises provide information technology-based goods and services largely, if not entirely, at a distance. Those working in such enterprises are likely to be employees living locally, but the customers and clients of the enterprise are located at a distance.

Realities

The telework mythology

Over and over again a limited number of examples of telework projects can be found in research literature from companies such as F. International, Rank Xerox and ICL in the United Kingdom, Integrata among others in Germany, and the Nykvarn Neighbourhood Centre in Sweden.

In the United Kingdom

There are only three organisations which use homeworkers in any number (Shirley, 1988, p. 26): F. International, Rank Xerox and ICL.

International was founded in 1962 by V.S. Shirley as one of the first teleworking organisations, and according to the founder it has "prospered to become the world's most successful model of this way of working". In the late 1980s, F. International had approximately 1 000 people of whom 200 were employed and 800 were freelance. All these staff work at a distance, *i.e.* from their homes, but also spend time at one of the company's small offices. At the end of 1986, 96 per cent of the workforce were women (*cf.* Shirley, 1985; Shirley, 1988; Schwohnke and Wicke, 1986; Kinsman, 1987; Maciejewski, 1987; Anderson, 1988; Huws, Korte and Robinson, 1990; ILO, 1990).

Rank Xerox International is a British company of which 51 per cent is owned by US-based Rank Xerox. In 1982, with the primary aim of reducing overhead costs, 21 sub-contractors were linked to the central London headquarters. In the late 1980s the number increased to 59. These sub-contractors are self-employed "networkers", working also for other companies (*cf.* Judkins, 1988; Schwohnke and Wicke, 1986; Kinsman, 1987; Maciejewski, 1987; Huws, Korte and Robinson, 1990).

The third British organisation using homeworkers in any number is ICL. Its homeworking scheme was set up in 1969, and in the late 1980s the company employed 300 female home-based teleworkers (*cf.* Shirley, 1988; Kinsman, 1987; Huws, Korte and Robinson, 1990).

In Germany

Integrata was founded in 1964, and introduced telework in 1983 when two women programmers with small children were provided with personal computers and modems in their homes to enable them to communicate remotely with the company. In 1984, the company also set up its first satellite office in order to better serve a major local client (*cf.* Heilman, 1988; Schwohnke and Wicke, 1986; Huws, Korte and Robinson, 1990).

A few other German companies (an insurance company, a translation agency, a typesetting company, a typing service company, a photo-typesetting company, and a scheme for disabled people) are reported to employ teleworkers (*cf.* Huws, Korte and Robinson, 1990).

In Sweden

Finally, the only known example of the so-called neighbourhood work centre should be mentioned: the Swedish Grannskapscentral in Nykvarn, some 25 miles from Stockholm. The Nordic Institute for Social Planning, NordPlan, initiated this project in 1982, and during its period of operation approximately ten workers, working for different companies, shared the computer and telecommunications facilities at the centre (*cf.* Bullinger, Fröschle and Klein, 1987; Mehlmann, 1988; Kinsman, 1987; Qvortrup, 1984; personal communication).

Measuring telework: estimates and problems

As stated above, this rather limited number of examples has been mentioned over and over again in the literature, sometimes almost as an excuse for the lack of more solid quantitative evidence. In this way a sort of mythology has been established. But what is really behind the examples?

Repeated optimistic predictions about the number of teleworkers have been put forward. However, if these predictions are compared with current statistics, the gap seems to be enormous.

According to Steinle, despite the vast potential for telework and the predictions "...actual applications of telework are rare. Estimates of numbers of teleworkers in Europe vary between a few thousand and over 100 000" (Steinle, 1988, p. 8). Similarly, in their critical book Schwohnke and Wicke conclude: "The current level of telework is much lower than is believed by supporters as well as opponents. On closer inspection, much-quoted examples are revealed to be single projects ranging from a handful to at most 100-200 teleworkers" (Schwohnke and Wicke, 1986, p. 102, my translation from German).

On the same lines, Korte (1988, p. 374) estimates that the incidence of telework in 1987 was:

Federal Republic of Germany	>1 000 teleworkers
France	<1 000 teleworkers
United Kingdom	approx. 3 000 teleworkers
Italy	<1 000 teleworkers
United States	approx. 10 000 teleworkers.

Based on these research findings, it is tempting to conclude that never have so many talked so much about so little. Or, as Steinle says: "There are more people doing research on telework than there are actual teleworkers" (Steinle, 1988, p. 8).

Before totally writing off telework, it is, however, important to emphasise that estimates of how many people are engaged in telework depend primarily on how it is defined. In his article "Telecommuting: The Trade-Offs of Home Work" (1989), Robert E. Kraut demonstrates that one of the basic problems in measuring telework is that researchers have based their quantitative analyses on the wrong categories. If "homeworkers" are defined as people who perform any income-producing work at home, then figures for the United States vary from 15 million to 23 million people, that is from 15 to 23 per cent of the civilian, non-farm labour force.

If more restrictive definitions of homework are used, estimates are scaled back dramatically. If "homeworkers" are defined as employees working for a company, using their own home as their principal place of work, then only approximately 1 million people, or 1 per cent of the non-farm, civilian labour force, in the United States are categorised as homeworkers. Recently, Link Resources Corp., a New York-based research firm, estimated that the number of US telecommuting employees in 1988 was 2.2 million, growing to 3.6 million in 1990 and to an expected 4.4 million in 1991 (*cf.* Mitch Betts, 1990).

Kraut concludes that "...disagreements about definition cause confusion about how much homework exists. Fifteen to 20 per cent of the non-farm labour force works at home at least some of the time, but less than 2 per cent does so for substantial parts of the work week" (Kraut, 1989, p. 22).

"Substitutors", self-employed, and "supplementers"

Based on his findings Kraut makes a distinction between three variants of homeworkers: substitutors, self-employed, and supplementers.

"Substitutors" are those who substitute work done at home for work done in a more conventional work setting. They are "telecommuters" in Nilles' terms, and they are the "teleworkers" measured by Steinle, Korte, and Huws, Korte and Robinson above. According to Kraut most researchers and commentators assume that teleworkers are primarily substitutors, *i.e.* employees of larger organisations who spend part of or all of their work week at home or at satellite offices or at shared-facility centres rather than in a traditional office. But this style of teleworking is more myth than reality. The two other styles of home-based employment are far more common: "self-employed", operating a home-based business, and "supplementers" bringing supplementary work home from their conventional office (Kraut, 1989, p. 23).

But why are teleworkers so often mistakenly identified as substitutors? Firstly, because interest in telework has been based on the potential substitution of electronic transportation for physical transportation, and the category of "substitutors" explicitly represents this interest. Secondly, because "substitutors" represent the visible teleworkers, while self-employed and supplementers are invisible teleworkers: working at home instead of going to the office, or sending their work results electronically to headquarters. But being self-employed, and using computers and telecommunications equipment to distribute one's services, is not inherently defined as telework. The same goes for supplementers or flexible work arrangements. For example, most university researchers work partly at home, or on trains and aeroplanes, using telephone and mail services, without defining themselves as teleworkers.

Kraut demonstrates that "self-employed" and "supplementers" are the two important telework categories. For example, data from the 1980 US census shows that a majority of work-at-home households were running their own businesses. Similarly, supplemental homework accounts for a number of categories of homeworker. Referring to a number of surveys, Kraut reports that: a majority of all employed managerial and professional specialty workers do some homework, but only occasionally and then to supplement rather than substitute for conventional office work. Over 80 per cent of teachers report some homework. Approximately half of all data-processing professionals work at home in addition to their regular work hours. University professors work part-time at home. A common reason for supplemental homework is that certain tasks require concentration that is unobtainable in a conventional office, while other activities are based on physical interaction: meetings, teaching, conversation *etc.*''In general, such employees used their conventional offices for social tasks and their homes for cognitive tasks'' (Kraut, 1989, p. 25).

Not only do these findings demonstrate that measuring telework is difficult, and that most surveys are based on very narrow categories; they also support the above-mentioned broad definition of telework as all work-related substitutions of telecommunications and related information technologies for travel.

Current telework schemes

In addition to measuring telework, it may be relevant to identify some trends from current telework schemes, and some models of telework organisations.

Recently, a number of publications have been emphasising the number of telework organisations. Based partly on an extensive survey of teleworkers and partly on available literature, *Telework: Towards the Elusive Office* by Huws, Korte and Robinson (1990) identifies 70 telework schemes in West Germany, France, the United Kingdom, Italy, the Netherlands, Sweden, and the United States employing a total of approximately 7 200 teleworkers (*cf.* Huws, Korte and Robinson, 1990, pp. 77-80). Similarly, the ILO journal *Conditions of Work Digest* in its topical issue *Telework* (Volume 9, N° 1, 1990) features 57 telework arrangements employing a total of approximately 5 600 teleworkers. Recently, a report written for the Commission of the European Communities has presented or updated five new telework schemes. And from Japan a few reports are available (*cf.* Fujino and Terashima, 1989; and Nakamura and Tsuboi, 1990).

Combining these publications, the following tentative list can be elaborated (Table 1):

Current trends

In the list 111 telework schemes are identified, employing a total of almost 12 000 workers. If we base our analysis of current trends in telework on the examples in the list – and it must be repeated that the list is not necessarily representative, being based only on published studies – the following interesting conclusions can be reported:

Firstly, it seems that most teleworkers (in the table approximately 75 per cent) are permanently employed. It seems that there is no correlation between telework and freelance arrangements, even though a number of examples can be reported.

Secondly, it seems that most teleworkers (again, approximately 75 per cent) are skilled workers (administration staff, managers, computer specialists, consultants, researchers, *etc.*). Even though there are examples of unskilled homeworkers, they represent a minority.

Thirdly, it seems that the dominant organisational models are flexible work arrangements (approximately 32 per cent) and satellite offices (approximately 33 per cent). Approximately 28 per cent are permanent homeworkers, while only a minority (7 per cent, most of whom are in pilot projects and publicly-funded experiments) work in shared facilities; however, some recent examples have provided us with promising results. In comparison, the analysis of Huws, Korte and Robinson shows that out of the identified 7 200 teleworkers, 2 800 are homeworkers (approx. 39 per cent), 2 300 mobile or flexible teleworkers (approx. 32 per cent), and 1 800 employed at satellite offices (approx. 25 per cent). Still, many small and longstanding telework arrangements are based on permanent electronic homework. Out of my 111 examples, 49 represent traditional homework arrangements.

Table 1. **Current telework schemes**

Name of organisation		Country	N° of workers	Work content	Organisation model	Date	Employement status	Comments
1	Credit Suisse	CH	60	skilled	sat. office	1987	per. emp.	comp. sp.
2	Stand. Tel. & Rad.	CH	5	skilled	sat. office	1985	per. emp.	comp. sp.
3	Printing Comp.	CH	4	skilled	sat. office	1987	per. emp.	experim.
4	Benglen	CH	6	skilled	shar. fac.	1985-86	per. emp.	
5	Text Compos. Comp.	D	9	unskill.	home	1970	self-emp.	
6	Printing Comp.	D	4	unskill.	home	?	self-emp.	
7	Chemical Comp.	D	3	skilled	home	?	freelance	transl.
8	Siemens	D	3	unskill.	home	1982	freelance	
9	State Gov. Baden-W	D	21	unskill.	home/sat.	1984-85	freelance	experim.
10	Integrata	D	20	skilled	home/sat.	1983	per. emp.	comp. sp.
11	ICR Neustadt	D	2	unskill.	home	?	?	
12	Programm. Serv.	D	80	skilled	sat. office	?	?	disabled
13	Lufthansa AG	D	50	unskill.	home	?	?	
14	GWK, Köln	D	20	unskill.	sat. office	?	?	disabled
15	ÖVA, Mannheim	D	300	skilled	flex/home	?	?	insur. ag.
16	Bausparkasse W.	D	800	skilled	sat/flex	?	?	insur. ag.
17	RWG Gmbh	D	5	skilled	home	?	?	
18	Tastsatz Lulay	D	15	unskill.	home	?	?	
19	Odental Texterf.	D	18	unskill.	home	?	?	
20	Bonner Übersetz.	D	5	mixed	sat/home	?	?	
21	Fotosatz Comp.	D	20	unskill.	home	?	?	disabled
22	Rehabilitationspr.	D	15	mixed	sat. office	?	?	
23	Reserach Comp.	D	2	skilled	flex.	?	?	
24	Le Monde	F	?	skilled	flex. w.a.	1987	per. emp.	journal.
25	France Télécom	F	14	unskill.	flex/home	1986	per. emp.	
26	Groupe PBS Télerg.	F	40	mixed	sat. office	1981	per. emp.	
27	Les Mutuelles Uni.	F	40	unskill.	flex/home	1984	per. emp.	
28	DEC, Valbonne	F	300	skilled	flex/home	1987	per. emp.	
29	Marne-La-Vallée	F	50	mixed	shar. fac.	1981	per. emp.	experim.
30	DGT-Pilot Project	F	14	mixed	home	?	?	disabled
31	Téléphone Service	F	37	unskill.	home	?	?	
32	Téléboutique	F	?	mixed	sat. office	?	?	
33	Researc Institute	F	1	skilled	home	?	?	disabled
34	Insurance Com.	F	700	skilled	sat. office.	?	?	insur. ag.
35	Insurance Com.	F	1	unskill.	home	?	?	
36	Bank	F	2	skilled	home	?	?	
37	Téléphone Mark	F	84	skilled	flex. w.a.	?	?	sales ag.
38	Sports Art. Comp.	F	15	skilled	flex. w.a.	?	?	sales ag.
39	Dataproc. Consult.	F	3	skilled	flex. w.a.	?	?	comp. sp.

40	The FI Group	GB	250	skilled	home/sat.	1962	per. emp.	
41	Rank Xerox Intern.	GB	800	skilled	home/sat.	1962	self-emp.	
42	Chamberlains	GB	55	skilled	home	1981	per. emp.	
		GB	75	mixed	flex/home	1981	self-emp.	
			15	mixed	flex/home	1981	self-emp.	
43	ICL	GB	320	skilled	flex/home	1969	per. emp.	
44	Hampshire Country	GB	20	skilled	flex w.a.	1987	per. emp.	
45	Typing Plus	GB	30	unskill.	home	1981	self-emp.	
46	Alternative Data	GB	4	skilled	home	1986	self-emp.	rural co.
47	Highland Data	GB	1	skilled	home	1973	self-emp.	rural co.
48	Geddes Lang. Serv.	GB	1	skilled	home	1985	self-emp.	rural co.
49	Tones Business EX.	GB	6	skilled	home	1989	Part-time	
50	Dep. Trade & Indu.	GB	69	skilled	home	1982	per. emp.	disabled
51	IT World Limited	GB	13	skilled	home	1988	per./free	disabled
52	Fintech	GB	6	skilled	flex/home	1984	freelance	journal.
53	Kent County Coun.	GB	2	mixed	home	?	per. emp.	
54	Systime	GB	150	skilled	home	?	?	
55	Nottingh. Build. S.	GB	1	skilled	home	?	?	
56	Olivetti Software	I	10	skilled	flex. w.a.	?	?	
57	ENI (Chem. Indu.)	I	1 000	skilled	flex. w.a.	?	?	
58	Univers. of Milan	I	50	skilled	flex. w.a.	?	?	experim.
59	O. Group	I	5	mixed	home	?	?	comp. sp.
60	Info. Services Co.	IRL	?	skilled	sat. office	?	?	off-sh.
61	Kao Company	J	550	unskill.	flex/home	?	per. emp.	
62	Kumamoto Prefect.	J	200	unskill.	shar. fac.	1988	per. emp.	experim.
63	Trading Comp.	J	16	skilled	home	1985	per. emp.	disabled
64	Shiki	J	4	unskill.	shar. fac.	1988-89	per. emp.	experim.
65	Mitsubishi Metal C	J	53	mixed	sat. office	1988	per. emp.	
66	Fujitsu Ltd.	J	11	mixed	sat. office	1979	mixed	
67	NEC Corporation	J	?	?	sat. office	1984	per. emp.	
68	Jamaica Dig. Inter.	JAM	?	?	shar. fac.	1988	per./free	off-sh.
69	Otto	NL	600	mixed	home	1981	tempor.	
70	Fokker Aircraft I.	NL	100	unskill.	sat. office	1978	per. emp.	
71	Fokus	NL	22	unskill.	home	1985-89	training	disabled
72	Min. of Traffic	NL	28	mixed	flex/home	1990	per. emp.	experim.
73	County of Gotland	S	30	mixed	sat. office	1987	per. emp.	
74	Selecta Contact AB	S	9	unskill.	sat. office	1988	per. emp.	
75	OK Petroleum AB	S	3	unskill.	sat. office	1990	per. emp.	
76	Nykvarn	S	10	mixed	shar. fac.	1982-84	per. emp.	
77	SIGA Serv. Centre	S	40	unskill.	shar. fac.	?	?	
78	US West	USA	3-600	skilled	home	1985	per. emp.	experim.
79	County of LA	USA	500	skilled	flex/home	1989	per. emp.	

Table 1. Current telework schemes (cont'd)

	Name of organisation	Country	N° of workers	Work content	Organisation model	Date	Employement status	Comments
80	State of Calif.	USA	230	skilled	flex/home	1988	per. emp.	
81	Travellers Insu. C.	USA	200	skilled	flex/home	1987	per. emp.	
82	Pacific Bell	USA	750	skilled	home/sat.	1985	per. emp.	
83	JC Penney	USA	200	unskill.	home	1981	tempor.	
84	SPAR	USA	20	unskill.	flex. w.a.	1980	freelance	
85	Wisconsin's Hosp.	USA	7	skilled	home	1984	per. emp.	
86	Control Data Corp.	USA	?	skilled	hom./sat.	1982	per. emp.	disabled
87	American Express	USA	10	skilled	home	1982	per. emp.	disabled
88	Lift, Inc.	USA	50	skilled	home	1973	per. emp.	disabled
89	IBM	USA	60	skilled	flex/home	1988	per. emp.	
90	US Army	USA	5	skilled	home	1980-81	per. emp.	experim.
91	Heights Inform.	USA	180	skilled	home	?	?	
92	Tymshare Cupert.	USA	40	skilled	flex. w.a.	?	?	
93	Interactive Syst.	USA	100	skilled	flex. w.a.	?	?	
94	Continent. I. Bank	USA	4	unskill.	home	?	?	
95	Blue Cross/Shield	USA	214	skilled	home	?	?	
96	American Airlines	USA	200	unskill.	home	?	?	
97	Satel. Data Corp.	USA	?	unskill.	sat. office	?	?	
98	Mountain Bell	USA	8	skilled	sat. office	?	?	
99	Hewlett P. Lab.	USA	1 000	skilled	home	?	?	experim.
100	Equitable Life	USA	5	skilled	home	?	?	after hours use
101	Blodgett Comp. S.	USA	100	unskill.	home	?	?	comp. sp.
102	Kemper Group	USA	?	unskill.	home	?	?	
103	S. Californ. Ass.	USA	15	mixed	home	?	?	
104	Major Color. Bank	USA	3	unskill.	home	?	?	
105	Aetna Life	USA	3	skilled	home	?	?	comp. sp.
106	Chase Manh. Bank	USA	4	skilled	home	?	?	comp. sp.
107	Ford Motor Comp.	USA	3	unskill.	home	?	?	
108	Investors' Serv.	USA	3	unskill.	home	?	?	
109	Manufs. Hann. Tr.	USA	4	skilled	home	?	?	comp. sp.
110	New York Tel.	USA	12	mixed	home	?	?	
111	Hawaii Telework	USA	16	skilled	shar. fac.	1989	per. emp.	

Abbreviations:

CH:	Switzerland	Home:	Homework
D:	Germany	Sat. office (or: Sat):	Satellite office
F:	France	Shar. fac.:	Shared facilities
GB:	Great Britain	Flex. w.a. (or: Flex):	Flexible work arrangements
I:	Italy	Per. emp.:	Permanently employed

IRL:	Ireland
J:	Japan
JAM:	Jamaica
NL:	The Netherlands
S:	Sweden
USA:	United States

Self-emp.:	Self-employed
Freelance (or: Free):	Freelancer
Tempor.:	Temporarily employed
Comp. sp.:	Computer specialist, programmer, etc.
Insur. ag:	Insurance agent
Off-sh.:	Off-shore company, providing cheap labour at distance
Experim.:	Experiment or pilot project, most often run or supported by public authorities
Transl.:	Translator
Journal.:	Journalist
Sales ag.:	Sales agent
Rural co.:	Rural community

Finally, it may be noted that some projects are particularly aimed at employing disabled people, and that the public sector is increasingly using telework as a way to organise its activities.

Examples of telework arrangements: unskilled, low-paid telework

Even though the trend is towards skilled teleworkers with flexible work arrangements, traditional unskilled and low-paid teleworkers still exist. One example is a German text composition company, established in the 1970s. Here, nine unskilled women work at home on a self-employed basis, paid piece rates which are among the lowest in the composing field, and suffering the problems of social isolation and overwork. The background to the arrangement is that the company is based in a rural area in which there are few job opportunities (cf. ILO, 1990, p. 79).

Even though it is different, Jamaica Digiport International (JDI) can be compared with the German text composition company. JDI was established in 1988 as a so-called "off-shore arrangement". It is a joint venture by American Telephone and Telegraph (AT&T) of the United States, Cable and Wireless of the United Kingdom, and Telecommunications of Jamaica. JDI provides advanced telecommunication technology to information processing firms in Jamaica's free zones. Currently, ten data processing businesses, employing approximately 600 people, are using the facilities, delivering data-entry work via satellite to companies and clients in the United States. However, the basic point of JDI is the low level of income in Jamaica. On average, a Jamaican doing skilled data-entry work earns between 10 and 20 per cent of the pay of a worker doing the same work in the United States (cf. ILO, 1990, p. 108ff).

These examples demonstrate that telework, being independent of distance, makes it possible to carry out work where it is cheapest, be it in less-favoured rural regions in the industrialised world or in Third World countries. A number of social issues are raised, but the fact that telework can provide urban areas with new forms of competition should also be considered.

My final example demonstrates what one might call the "modern" type of telework. In the United Kingdom, the consultancy company Chamberlains was established in 1981. Since then the firm has grown rapidly on a networking model. The company provides consultancy on a range of personnel services. There are a central office and four regional offices with approximately 25 permanently employed skilled and unskilled staff. In addition, there are 75 so-called "permanent temporaries" who are employed by Chamberlains but carry out their work on clients' premises, providing a flexible workforce for a number of different tasks, and a number of "licensees" aod "associates". The flexible homeworkers are equipped with personal computers, accessing the central office by telephone, fax and electronic mail. It seems that such networking models will be more and more typical during the years to come, influencing traffic patterns, the need for decentralised public services, lifestyles, etc.

Satellite work centres

It has been much debated whether the type of unskilled electronic homework exemplified above is acceptable. The employees are isolated from the working partnership of a traditional office, and quite often the public services (childcare etc.) available are not satisfactory. On the other hand the flexible working hours and the reduction of transportation costs and time are often appreciated.

One possibility for a company is to establish a satellite centre. Fokker Aircraft Industries provides an example. Its main production facilities are at Schiphol Airport. A data-entry satellite centre was opened in Alphen aan de Rijn in 1978. The centre was established in a residential area because of problems of attracting part-time data-entry workers at the offices in Schiphol. Today 22 employees, all women, work in the satellite centre – the equivalent of 17.5 full-time jobs – entering data from the many divisions of Fokker on computer terminals. Once a day the data are transmitted from the satellite centre to the different main-frame computers in the offices in Schiphol and Amsterdam. Jan Fokkema of Werkgroep'2 Duizend judges that the project has been successful for both the company and the teleworkers. For more than 12 years, the company has had a stable pool of experienced data-entry typists, while the employees are satisfied with having a well-paid job close to home (cf. ILO, 1990, p. 85ff; Fokkema, 1990; personal communication).

The role of city authorities in relation to satellite work centre arrangements is limited. However, such arrangements can be justified in general city planning schemes, allowing for centres to be established in residential areas. Such initiatives should be accompanied by provision of relevant public services in the same areas. In this way some of the problems of telework for unskilled workers can be avoided.

Shared-facility centres

While satellite centres are established by private companies, typically relocating an administrative department, public authorities can support the building of shared-facility centres. A number of such centres have been launched during the last 10 years, starting with the French Marne-la-Vallée centre in 1981 and the Swedish Nykvarn neighbourhood centre in 1982, and continuing with the Swiss neighbourhood centre in Benglen, close to Zürich, and the recent Hawaii Telework Demonstration Project.

In a small shared-facility centre, a "neighbourhood centre", people can work for different companies using the shared and publicly subsidised facilities. One such example is the Swedish "Grannskap" centre. A number of difficulties had to be overcome before the project could get started: lack of interest on the part of the local authorities; union anxiety that projects of this kind would undermine shop-floor solidarity; and managerial scepticism based on the possibility of a reduction in profits resulting from the extra (not insubstantial) capital outlay. Nevertheless, supported by the Nordic Institute for Social Planning and sponsored by the public authorities and by various firms, the centre was established in November 1982 in the small community of Nykvarn (population 5 000), some 30 km from Stockholm. There were approximately ten workers at the "grannskapscentral". During the project satisfaction was unanimously expressed about the reduction in commuting time, and about the fact that the new office was in the immediate vicinity of the workers' homes. Satisfaction was expressed, too, about the fact that the neighbourhood centre dispelled the isolation normally endemic to remote terminal work, in a way which was generally considered to be socially acceptable. But dissatisfaction was expressed about the fact that one now had to go to work in two places: one's former office, where social contact with colleagues suffered from the relative infrequency of visits, and the neighbourhood centre, where, because so many different professions were represented, social contact was mainly centred on leisure activities. In the long run, the negative aspects became dominant.

The dissatisfaction about the lack of social contact with the main offices resulted in low utilisation of the centre, which meant that the social fellowship of the centre gradually disappeared. In July 1985 the project was abandoned. Similar problems were experienced in the French Marne-la-Vallée project (Qvortrup, 1984, p. 164ff; Engström et al., 1985; ILO, 1990, p. 135ff).

Fundamentally, however, I think that the problems were related to social innovation. In the early 1980s the concept of telework was little known, and fears about the social and organisational impact influenced the participation of workers, employers, unions and local authorities. Recent experiments seem to demonstrate that attitudes have now changed.

One example is provided by the Hawaii Telework Demonstration Project which was formally opened on July 14, 1989. It was funded by a $125 000 appropriation of the Hawaii State Legislature with matching contributions from local businesses. The centre is located in the Mililani High Tech Park, at the edge of a major new housing development, some 15 miles from downtown Honolulu. The centre has 16 workspaces and one space for an on-site administrator. Eight of the teleworkers are public employees while eight others are private business employees; all employees live in the nearby community. These teleworker employees engage in tax, legal, planning, programming and related information-intensive tasks. In more than a year of operation the demonstration project has been highly successful. Commuting costs of teleworkers decreased more than 80 per cent. Both supervisors and workers report job productivity either stable or improved. All teleworkers involved report that job satisfaction increased significantly, as did quality of life in general. Based on these results, it is expected that a few hundred telework centres will be opened, distributed across all of the inhabited Hawaiian Islands (Harms, 1990; Information Times, Vol. 2, N° 9, 1990; Hawaii Telework Center Demonstration Project, 1990).

Community teleservice centres

My last example of shared-facility centres is the Nordic Community Teleservice Centres (CTSCs). A community teleservice centre may be defined as a centre where information technology is placed at the disposal of the citizens of a specific local community within a marginal geographical location, so that communal use may be made of the facilities. The purpose of the CTSC is to counteract geographically determined disadvantages which affect the local community, whether of an economic, educational or cultural nature, and whether those disadvantages concern employment, services, or other infrastructure facilities. The first centres were established in Denmark and in Sweden in 1985, and today more than 50 community teleservice centres are operating in Denmark, Norway, Sweden, Finland, Scotland, Ireland and Canada.

In a number of cases work stations have been established for unskilled distance working at the teleservice centre. The idea was that distance working conducted at the teleservice centre as a communal activity would eliminate the problems of distance working in the traditional sense of "home working", which isolated the employee from a working community.

Even though distance working at a communal centre is better than home working, the idea has failed. Firstly, because firms in the urban centres, other things being equal, seem to prefer local, unskilled workers to unskilled workers at a remote teleservice centre. Secondly, because even though the problem of isolation was eliminated, the concept of distance working based in a teleservice centre still reinforced the traditional periphery-centre dichotomy between unqualified jobs in the rural regions and qualified jobs in the town. A similar example is found in Sweden, where a remote work centre was set up in the far north of the country in a region of high unemployment to provide back-office functions for Stockholm-based organisations. However, the jobs created proved to be low-level and routine, and when there was a fall in demand it was these workers who were laid off. They simply could not compete with similar workers living in the same region as the central office (Huws, Korte and Robinson, 1990, p. 214).

Consequently, it has been a general conclusion that the only type of distance working normally carried out at a teleservice centre should be competence-based work such as architecture, consultancy, research, administration, *etc.*, executed by people who by using the facilities at the centre avoid having to go to town every day (Qvortrup, 1989; Hetland and Olsen, 1988; Hetland, Knutzen, Meissner, Olsen, 1989). Recently, such work places have been established in centres in Northern Norway and in Northern Scotland, and the first findings from these projects seem to prove that competence-based work from these rural centres or from home, profiting from the (library, database and distance educational) facilities available in the nearby community centre for urban based clients, can be successful (Review of Current Experiences and Prospects for Teleworking, 1990; Newsletter on Community Teleservice Centres, 1990).

Public sector telework schemes

The most obvious way to support telework is of course to employ teleworkers within the public sector. One such example is provided by the County of Los Angeles, California. The county has a population of 8.5 million, and it employs 80 000 workers in 39 different departments, including justice, public safety, health services, social services, parks and recreation, *etc.* A plan to implement telework for county employees was adopted by the County Board of Supervisors in April 1989 in order to save costs on office space, improve the recruitment and retention of employees, limit traffic congestion, improve flexibility and reduce rates of absenteeism and sick leave. From the employees' perspective, telework was viewed as providing a means to reduce their commuting time and expenses and to increase their flexibility in handling work and family schedules.

In the first phase 300 employees from 15 different departments joined the programme, enjoying flexible work arrangements, working at home between one and four days a week; and the number of flexible teleworkers is growing. The project is looked upon with enthusiasm and it is expected that 2 000 of the county's 17 000 employees in downtown Los Angeles could be working at home in the next five years (ILO, 1990, p. 52ff. Review of Current Experiences and Prospects for Teleworking, 1990).

Individual impact of telework

Individual, organisational and societal impact of telework

Normally, advantages and disadvantages of telework are analysed on three levels:
- impact on the individual employee;
- impact on the organisation;
- impact on society at large.

This section will summarise the impact of telework on the individual teleworker.

Impact of telework on the individual teleworker

As telework is organised in many different ways, some being carried out in the context of a shared-facility centre and some being totally home-based, and as work content varies from monotonous low-skilled clerical word-processing to diversified, high-skilled and specialised problem-solving, it is difficult to put forward any universal statement about the human aspects of telework.

Yet, a number of ways telework impacts on the individual teleworker can be listed:

The need for flexibility seems to be the dominant reason for telework (*cf.* Huws, Korte and Robinson, 1990, p. 60; Olson, 1988, p. 81; Korte, 1988, p. 385; Kraut, 1989, p. 34). Telework creates a greater flexibility in the distribution of working time, and the teleworker has a sense of autonomy in his or her organisation of the working day. But is flexibility an advantage or a disadvantage? M.H. Olson reports that 70 per cent of the people emphasising flexibility as an important aspect of telework were women, 50 per cent of them mothers of small children. Similarly, in the Empirica survey 73 per cent of the teleworkers surveyed were female. In this survey only 30 per cent of male teleworkers rated the childcare advantage as important, while 80 per cent of female teleworkers rated the childcare advantage as ''important'' or ''very important'' (Olson, 1988, p. 81; Huws, Korte and Robinson, 1990, p. 104).

However, it is very often difficult to make an unambiguous statement. In a 1982 UK survey, one of the respondents, when asked what was the main advantage of working from home, replied ''being with the children all day'', but when asked the main disadvantage of telework gave the same reply (Huws, Korte and Robinson, 1990, p. 59). It is the general experience that the homogenisation of work and leisure time has led to complaints stemming from role conflicts: the teleworker is never ''really at home'', nor ''really at work'' (Connolly, 1988, p. 9).

To summarise the impact of telework on individuals it is remarkable that experiences are very different, and that they are divided strictly in accordance with two major types of teleworkers: the low-skilled and low-paid, typically female clerical worker and the well-educated, typically male managerial or self-employed worker. The experience of the word-processing type of teleworker arouses fears that telework will lead to increased isolation and reduced influence for the worker, sitting all alone (except for the children) behind her domestic terminal. Deprived of professional and social intercourse with her colleagues, burdened by such domestic duties as child-minding and housework while she is trying to get on with the job, and constantly electronically monitored by her employer, her ''emancipation'' from the office will be merely synonymous with a return to nineteenth-century homework. On the other hand, there are those who regard remote terminal office work in a more positive light: no more transport problems; flexible working hours; familiar and personally created surroundings; no interruptions from telephone calls and office colleagues. Normally, however, the latter points of view are expressed by male, highly-skilled managerial or self-employed teleworkers (*cf.* Qvortrup, 1984, p. 165).

In conclusion, one should not generalise about the situation of the individual teleworker. The impact depends on the specific type of telework. Similarly, one should not suggest any universal regulations of telework. It may well be that the routine word-processing type of electronic homework should be avoided, while highly qualified managers or self-employed consultants should benefit from their socially isolated part-time homework, being able to concentrate uninterrupted on their ''cognitive tasks'' (Kraut, 1989, p. 25).

Similarly, it may well be that shared-facility centres or satellite offices are relevant solutions to the problems of the word-processing type of teleworker, because such organisations combine flexibility and reduced commuting with companionship at work, while part-time homework or flexible work arrangements seem to be the optimal solutions for the educated and skilled teleworker and/or for the teleworking manager. Along the same lines, two recent surveys from Singapore conclude that while the work arrangement of a full-time staff, working partly at home and partly in the office, is acceptable to a majority of both computer managers and computer professionals, full-time telework is not acceptable to either of them (Yap, Raman and Lim, 1990).

Organisational impact of telework

Organisational barriers

From a technical point of view, a majority of the workforce could work at home or in local telework centres. However, the number of teleworkers continues to be limited. So the question is not why telework exists, but why it is so infrequent.

According to advocates of telework, most organisations and firms can benefit from it: they can save money on location costs when some tasks are carried out in less costly facilities or in employees' homes, say Cross and Raizman. In addition, some teleworking projects have shown that productivity of information workers increases significantly when they are moved from the central office to their home-based work stations. And finally, a great deal of money can be saved if companies extend their remote operations beyond national boundaries, exploiting low-paid information workers in Third World countries. Cross and Raizman report that the North American Satellite Data Corporation "...has telecommuters working in the Caribbean for a fraction of New York City wages and is considering expanding its programme to India" (Cross and Raizman, 1986, p. 9).

In spite of all these business-related advantages, many researchers believe that present organisational structures and managerial attitudes form a major barrier to the rapid spread of telework. American surveys demonstrate that only a small minority of managers are in favour of telework, while a majority say that they need the personal interaction of office work to make employees work effectively (Huws, Korte, Robinson, 1990, p. 28). In 1988 M.H. Olson stated that "the one overlooked social force which is and will continue to be a major barrier to telework is organisational culture, and in particular management style" (Olson, M.H., 1988, p. 97f). Similarly, Empirica's European survey confirmed "conservative attitudes and institutional inertia... as the major factor currently preventing European companies from making use of telework" (Huws, Korte, Robinson, 1990, p. 29), and recently the two above-mentioned surveys of attitudes to telework among female computer professionals and computer managers in Singapore concluded that while 73 per cent of the computer professionals were in favour of telework, only 21 per cent of the managers indicated that they were "likely" or "very likely" to adopt telecommuting for computer professionals under their supervision (Yap, Raman and Lim, 1990). The employees seem to be more progressive than their managers.

Is managerial scepticism based only on conservativism, that is on psychological backwardness? No, according to Kraut, at least partly, the scepticism is rooted in realities. The reason why so few employers have converted conventional office arrangements into satellite offices or homework is that informal communication based on physical proximity is one of the most important tools for organisational coherence. Informal communication serves many functions in organisations. It is "...the basis of supervision, socialisation, social support, on-the-job learning by doing, and of the reproduction of corporate know-how and culture. Moreover, the informal communication among co-workers helps provide the major satisfaction denied to homeworkers – socialising and friendly interaction" (Kraut, 1989, p. 26). On the same lines, Connolly summarises the managerial attitude as follows: "The major problem with telecommuting boils down to the fact that a company can't teach values through a computer terminal."

Management tends to be distrustful of employees it can no longer see (Connolly, 1988, p. 8 and p. 9).

Trade unions' view of telework

Traditionally, trade unions have been sceptical about telework. One good example is provided by Judith Gregory of the North American Department for Professional Employees, AFL-CIO. According to her, in the United States in 1985, 98 per cent of all secretaries were women, and females account for the overwhelming majority of clerical workers. For Judith Gregory, telework does not represent a gain in flexibility. On the contrary, telework for unskilled female workers is a step backwards to a twentieth century equivalent of nineteenth century cottage industry, undermining pay levels, job security, quality of work, childcare services and occupational health and safety standards, and reinforcing part-time employment and social isolation, thus counteracting the ability of female clerical workers to organise on their own behalf (Gregory, J., 1985). Consequently, a 1983 resolution on computer homework adopted by the Constitutional Convention of the AFL-CIO union ''...resolved that the AFL-CIO calls for an early ban on computer homework'' (National Research Council, 1985, p. 153).

Similarly, English writer Zimmerman focuses on the tendency of telework to intensify already apparent distinctions between more prestigious, executive or self-employed work and menial, low-paid information work. ''Telecommuting promises two very different types of work experience for those at the upper and lower ends of the occupational scale: data-entry clerks and secretaries will handle routine tasks under continuous computer scrutiny of their performance and hours, while professionals will have discretionary working hours and unrestricted freedom to use computers for personal tasks, such as home accounting and database access'' (Zimmerman, 1986, p. 30).

Along the same lines, in West Germany Schwohnke and Wicke summarise their critical book on telework by saying that telework is a kind of work which ''damages the personality and increases the profit''. Consequently, ''by all means it must be stopped'' (Schwohnke and Wicke, 1986, p. 220, my translation from German).

Recently, however, trade unions seem to have modified their resistance to telework. For example, in 1985 the biggest Danish union for clerical workers decided under no circumstances to accept electronic homework (cf. Vedel, 1987, p. 62f). However, in its 1989 ''declaration of principles'', the same union agreed to maintain its resistance to electronic homework, but only ''as to principles''. ''If homework becomes established,'' the declaration says, ''the union will follow the development carefully and see that all existing agreements regarding wages and working conditions are observed'' (HK, 1989, my translation, my emphasis).

Conclusion

During the 1980s there has been much scepticism about telework on the part of employers as well as employees and their trade unions. However, in the first place scepticism has focused on the substitution model of telework, and during the period flexible work arrangements and self-employed teleworkers have gradually emerged, thus undermining the general resistance to telework. Secondly, on both sides of the labour market the current attitude seems to be much more pragmatic. One should not talk about telework as such; rather, one should specify the different ways of organising telework: some are considered to be acceptable, perhaps even beneficial, others are not.

Societal impact of telework

In this section, we will summarise the impact of telework on society at large, including the city as a totality. Telework will influence the transportation pattern, and the total design of our future cities.

The telecommunications-transportation trade-off

From the very beginning of telework-related research it has been argued that telework will help ease city traffic problems and that it will also provide society with an energy-use advantage (Cross and Raizman, 1986, p. 12), and as early as 1974 the first analysis of the telecommunications-transportation trade-off was elaborated (*cf.* Nilles, Carlson, Gray and Hanneman, 1976).

Still very little is still known about the substitution of telecommunications for travel. First of all, telecommuting is sparsely scattered in society, making quantitative analysis very difficult to verify. Secondly, many organisations that may have large numbers of part-time home teleworkers are not aware of the fact, just as sole proprietors of home-based businesses are not aware of being teleworkers. Thirdly, telework offers very different travel reduction options: complete substitution by a few full-time homeworkers; partial substitution by flexible and mobile teleworkers; reduction of the length of the journey to work for teleworkers associated with shared-facility centres; or a shift of commuting away from rush hours. Fourthly, the problems over the definition of telework make it difficult to compare the few existing telecommunications-transportation trade-off analyses.

In 1988 Jack Nilles concluded that because of the lack of data it was not possible to go beyond speculation as to the future of transportation. However, a conservative estimate based on a 1985-1986 survey, using the Blackman version of the technological substitution curve (Linstone and Sahal eds., 1976) forecasts that in 1990 out of 125 million workers in the United States, of whom 70 million are information workers, 1.2 per cent (800 000) of information workers will be teleworkers, and 0.8 per cent (500 000) will be homeworkers. In 2005 there will be 140 million workers, of whom 82 million will be information workers; 9.4 per cent (8 million) will be teleworkers and 7.4 per cent (6 million) homeworkers.

Thus, Nilles summarises, "transportation planners could easily rationalise postponing considerations of telecommuting as a means of altering transportation patterns... On the other hand, the transportation congestion problems in most cities continue to mount... Hence, one could also take the approach that telecommuting should be at least as thoroughly and widely tested as many of our other approaches to congestion relief – if only to get good data sooner" (Nilles, 1988, p. 309).

Additional analyses of the impact of telework on regional and urban development can be found in Giaoutzi, M. and Nijkamp, P. (1988) and in Akademie für Raumforschung und Landesplanung (1987), *cf.* especially Dostal, W. (1987).

The post-modern city

However important the future of transportation, one should not pretend that the city of tomorrow will be like the city of today but with altered transportation patterns and fewer traffic jams. If 10 per cent of information workers become teleworkers, this will not only change transportation policies, it will influence the very identity of our cities.

Even though most researchers have restricted themselves to the repetition of beloved, mythological telework projects, to statistical opinion polls on telework, or to quantitative traffic forecasts, a few scientists have ventured into qualitative social forecasting (see Bell, 1973, pp. 3-45 on the theories and problems of social forecasting). Two of those are Maria Christina Gibelli and Mark Hepworth.

As I emphasised in the introductory section, the crucial question is: with little traditional trading, with no big industry, and with a decreasing amount of traditional, centralised bureaucratic administration, what will actually be the functional basis, the infrastructural nervous system, and the architectural body of the future, "post-modern" city?

At the OECD experts meeting in Paris in 1987, Maria Christina Gibelli described two contradictory tendencies in the development of the city:

- On the one hand she identified the "Fordist Metropolis", characterised by a strict relationship between form and function, by specialised zones of industry, administration and residential quarters, with a cultural-symbolic centre, and with a regular pulse, beating like a heart, everyday sending flows of commuters from the outskirts into the centre and back again.
- On the other hand she put forward the concept of the "Postmodern Metropolis", heavily reliant on "network co-operation", less rigidly structured, capable of hosting changing functions in the same buildings, thus losing its unambiguous architectural signals. In order to host flexible information-based activities, traditional architectural typologies clearly related to specific functions will lose their significance in favour of flexibility and multi-purpose adaptability. Consequently we witness a trend away from traditional architecture saying that "this is an office", "this is a railway station", "this is a museum", to a post-modern or "negative" architecture saying that "this is not an office" (because offices have become invisible), and "this is not a railway station" (because railway stations are used as museums).

Based on networking and thus partly eliminating the meaning of geographical distance, this post-modern city will mix up office work quarters, shopping centres, and residential quarters, with the "intelligent home" as the symbolic node of home-working, home-shopping, home-banking, and home-based leisure.

Such Utopian – or dystopian – forecastings are further elaborated by Robins and Hepworth (1988) and by Hepworth (1989). In the post-modern city, they forecast, "...the co-ordinates of spatial orientation and identity will be the confines of the 'electronic home' on the one hand, and, on the other, what Frederick Jameson calls the 'post-modern hyperspace of the global village'" (Jameson, 1984, Robins and Hepworth, 1988, p. 192). This is precisely one of the basic, qualitative types of impact of telework: I am "at my job" even though I am "at home". I am part of an "online community" sitting solely behind my work station.

CONCLUSIONS AND RECOMMENDATIONS

Information and demonstration projects

There are so many myths surrounding telework that very basic information is needed about the concept and its potential and its disadvantages. Traditionally, telework has been defined as work at a site away from the company's central workplace; the results of the work are then communicated electronically to the employer or to the central office. Many examples show that this definition is too narrow. Most telework is based on flexible work arrangements or work by the self-employed at their own, home-based businesses using computers and telecommunications facilities to interact with subscribers, contractors and colleagues. In order to disseminate knowledge about telework, I would recommend public information campaigns, especially encouraging city authorities to set up projects to support telework.

Quantitative forecasts

After a period of heady optimism, today the dominant attitude seems to be that telework has no future. In relation to such attitudes it is important to emphasize that telework is likely to grow steadily, albeit more slowly than early commentators predicted (Huws, Korte and Robinson, 1990, p. 207). In addition, the growth of telework seems to lie mainly in flexible work arrangements, networking and self-employment.

Research

Traditionally, research on telework has focused on teleworkers according to the traditional, narrow definition, that is on those who substitute work done at home for work done at a central office. However, evidence shows that so-called "supplementers" (those bringing supplementary work home from their conventional office) and "self-employees" (those operating a home-based business) are far more common (cf. Kraut, 1989). Consequently, analyses, research projects and large-scale experiments evaluating attitudes to and impact of telework based on the broad definition are much needed.

Urban planners should differentiate between different categories of telework

The division between "traditional" teleworkers (the substitutors) and "new" teleworkers (the supplementers and the self-employed) seems to reflect the division between unskilled, clerical office telework and the skilled work of managers, researchers, consultants, independent information providers, etc. There is still much well-founded scepticism on the part of employers and employees and their unions about telework in the traditional sense. This scepticism should be taken into consideration; but "traditional" telework should not be confused with other and more acceptable types of telework.

Urban planners' attitude to unskilled telework

In the context of urban planning, it should be decided whether traditional, substitutional telework should be supported or avoided. If it is accepted, it is important to reduce the negative individual and organisational impact of such telework, for example by offering access to shared-facility centres and by offering relevant public services (schools, libraries, social security offices) and private services (shops, banks, cafeterias) in the neighbourhood of the shared-facility centres.

Urban planners' attitude to skilled telework

It must be appreciated that the number of teleworkers using flexible work arrangements, and the number of self-employed telework businesses, are growing rapidly. For such arrangements, facilities for access to advanced information and telecommunications services should be offered; in addition, public and private services are needed within residential areas. Furthermore, in the long term the architectural design of private housing and of residential areas should reflect these trends.

Telework: a symbol of organisational change

The qualitative and long-term impact of telework should be considered at the organisational level. The traditional concept of the workplace as a fixed geographical space will become increasingly outmoded, and will be replaced by more abstract notions of "online communities", "networking", "intellectual spaces", *etc.* Of course, such trends should also be reflected by the planning and organisation of future public administration at the urban level. Firstly, by decentralising those services regularly accessed by the public. Secondly, by critically examining the basic organisational philosophy of current public administration.

Preventive traffic planning is needed

Traditionally, urban planning has tried to remedy, rather than to prevent, traffic congestion problems, and information technology is still used for traffic planning purposes, to cure the patient rather than to prevent the malady. With telework, a unique opportunity for prevention is presented: in the long term, transportation structures can be modified and peak-hour problems reduced. Hence, telework should be at least as thoroughly and widely tested as many of our other approaches to congestion relief (*cf.* Nilles, 1988).

Long-term thinking is needed

Finally, the qualitative and long-term impact of telework at the societal level should also be considered. In the long term cities will become increasingly based on networking, and today's geographically separated functional quarters will be mixed up, with the "intelligent home" as the symbolic node of work, shopping, banking, leisure, and social interaction. Consequently, the co-ordinates of spatial orientation will be radically modified. As cities are inert societal arrangements, such long-term trends should be reflected by today's urban planning: what will be the functional basis, the infrastructural nervous system, and the architectural form of the future, "post-modern" city? I don't believe in the "disappearance of the city", but I do believe that we have to modify our traditional urban planning concepts, including information networking paradigms in addition to traditional urban planning paradigms.

BIBLIOGRAPHY

AFL-CIO (1985), "Resolution on Computer Homework", in National Research Council, *Office Workstations in the Home,* National Academy Press, Washington, D.C., p. 152f (originally, the resolution was adopted in 1983).

AKADEMIE FÜR RAUMFORSCHUNG UND LANDESPLANUNG (eds.), (1987), *Räumliche Wirkungen der Telematik,* Curt R. Vincentz Verlag, Hannover.

ANDERSEN, J. (1988), "'New Homeworking' in Social Context", in *Concerning Home Telematics* (Rijn, F.V. and Williams, R. eds.), North-Holland, Amsterdam, pp. 423-438.

BELL, D. (1973), *The Coming of Post-Industrial Society. A Venture in Social Forecasting,* Basic Books, New York.

BELL, D. (1979), "The Social Framework of the Information Society", in *The Computer Age: A Twenty-Year Perspective* (Dertouzos, M.L. and Moses, J. eds.), The MIT Press, Cambridge, Massachusetts.

BETTS, M. (1990), "Bush to Push Work-at-Home", in *Computerworld,* Nov. 26.

BULLINGER, H.-J, FRÖSCHLE H.-P. and KLAIN, B. (1987), *Telearbeit: Schaffung dexentraler Arbeitsplätze unter Einsatz von Teletex,* AIT, Hallbergmoos.

CHRISTENSEN, K.E. (ed.), (1988), *The New Era of Home-Based Work. Directions and Policies,* Westview Press, Boulder, Colorado, and London.

CONNOLLY, S. (1988), "Homeworking Through New Technology: Opportunities and Opposition", Parts 1 and 2, in *Industrial Management & Data Systems,* Sept./Oct. pp. 3-8 and Nov./Dec. pp. 7-12.

CRAIPEAU, S. and MAROT, J.-C. (1984), *Telework: The Impact on Living and Working Conditions,* European Foundation for the Improvement of Living and Working Conditions.

CROSS, T.B. and RAIZMAN, M. (1986), *Telecommuting, The Future Technology of Work,* Dow Jones-Irwin, Homewood, Illinois.

DOSTAL, W. (1987), "Telearbeit. Phänomen, Bedeutung, Probleme", in *Räumliche Wirkungen der Telematik,* Vincentz Verlag, Hannover, pp. 111-133.

EMPIRICA (1986), *Trends and Prospects of Electronic Home Working. Results of a Survey in Four Major European Countries,* Commission of the European Communities, Luxembourg.

ENGSTRÖM, M.-G., PAAVONEN, H. and SAHLBERG, B. (1985), *Grannskap 90. Närarbete pa distans i informationssamhället,* Teldok, Stockholm.

FOKKEMA, J. (1990), *Telewerken Dichterbij?* Werkgroep'2 Duizend, Amersfoort.

FORESTER, T. (ed.), (1980), *The Microelectronics Revolution,* Basil Blackwell, Oxford.

FUJINO, K. and TERASHIMA, T. (1989), "A New Work Environment: Applied Advanced Telecommunication Facilities – C&C Satellite Office", in *Proceedings of the Pacific Telecommunications Conference "89,* Hawaii, pp. 423-431.

GIAOUTZI, M. and NIJKAMP, P. (eds.), (1988), *Informatics and Regional Development,* Avebury, Gower Publishing Company, Aldershot.

GIBELLI, M.C. (1988), "Urban Planning Strategies and Tools to Cope with Technological and Socio-Economic Change in Metropolitan Areas", in *Urban Development and Impacts of Technological, Economic and Socio-Demographic Changes,* OECD, pp. 88-104.

GREGORY, J. (1985), "Clerical Workers and New Office Technologies", in National Research Council, *Office Workstations in the Home,* National Academy Press, Washington, D.C., pp. 112-124.

HARMS, L.S. (1990a), "Home-Based Telework in Hawaii?" Department of Communication Telehomework Workshop, University of Hawaii at Manoa, Hawaii.

HARMS, L.S. (1990b), *Telework in Sociotechnical Networks: Networking by Teleprofessionals,* unpublished paper, Department of Communication, University of Hawaii at Manoa, Hawaii.

HEILMANN, W. (1988), "The Organizational Development of Teleprogramming", in *Telework: Present Situation and Future Development of a New Form of Work Organization* (Korte, W.B., Robinson, S. and Steinle, W.J. eds.), North-Holland, Amsterdam, pp. 39-58.

HEPWORTH, M.E. (1989), *Geography of the Information Economy,* Belhaven Press, London.

HETLAND, P. and Olsen, O.E. (1988), "Hvor langt er det til fjernarbeids-markedet?", in *Distansarbete i Norden* (Berg, A.-J., Lie, M., Hetland, P., Olsen, O.E. and Kuusi, O. eds.), Nordrefo, Helsingfors, pp. 59-79.

HETLAND, P., KNUTZEN, P., MEISSNER, R. and OLSEN, O.E. (1989), *Naert, men likevel fjernt. Telematikk och lokal utvikling,* Rogalandsforskning, Stavanger.

HILTZ, S.R. (1981), *The Impact of a Computerized Conferencing System on Scientific Research Communities,* Newark, N.J.: Computerized Conferencing and Communications Center, New Jersey Institute of Technology, Research Report N° 15.

HILTZ, S.R. (1984), *Online Communities. A Case Study of the Office of the Future,* Ablex Publishing Corporation.

HUWS, U. (1988), "Remote Possibilities: Some Difficulties in the Analysis and Quantification of Telework in the UK", in *Telework: Present Situation and Future Development of a New Form of Work Organization* (Korte, W.B., Robinson, S. and Steinle, W.J. eds.), North-Holland, Amsterdam, pp. 61-76.

HUWS, U., KORTE, W.B. and ROBINSON, S. (1990), *Telework: Towards the Elusive Office,* John Wiley & Sons, Chichester.

ILO (1990), *Telework* (ed. Jankanish, M.B.), *Conditions of Work Digest,* Vol. 9, 1, International Labour Office, Geneva.

INFORMATION TIMES, INC. (1990), *Information Times,* Vol. 2, N° 9, Honolulu.

JAMESON, F. (1984), "Postmodernism, or the Cultural Logic of Late Capitalism", in *New Left Review,* N° 146, pp. 53-92.

JUDKINS, P.E. (1988), "Towards New Patterns of Work", in *Telework: Present Situation and Future Development of a New Form of Work Organization* (Korte, W.B., Robinson, S. and Steinle, W.J. eds.), North-Holland, Amsterdam, pp. 33-38.

KORTE, W.B. (1988a), "Telework. Potential and Reasons for its Utilization from the Organization's as well as the Individual's Perspective", in *Concerning Home Telematics* (Rijn, F.V. and Williams, R. eds.), North-Holland, Amsterdam, pp. 373-392.

KORTE, W.B. ROBINSON, S. and STEINLE, W.J. (eds.), (1988b), *Telework: Present Situation and Future Development of a New Form of Work Organization,* North-Holland, Amsterdam.

KRAUT, R.E. (1989), "Telecommuting: The Trade-Offs of Home Work", in *Journal of Communication 39*[3], pp. 19-47.

LANCASTER, F.W. (1982), *Libraries and Librarians in an Age of Electronics,* Information Resources Press, Arlington, Virginia.

LEBEVRE, B. (1977), "The Impact of Microelectronics on Town Planning", in *Impacts of Science on Society,* Vol. 27, N° 2; also published in *The Forester,* 1980, pp. 473-487.

LINSTONE, H.A. and SAHAL, D. (eds.), (1976), *Technological Substitution: Forecasting Techniques and Applications,* Elsevier, New York.

MEHLMANN, M. (1988), "Social Aspects of Telework: Facts, Hopes, Fears, Ideas", in *Telework: Present Situation and Future Development of a New Form of Work Organization* (Korte, W.B., Robinson, S. and Steinle, W.J. eds.), North-Holland, Amsterdam, pp. 101-110.

NAKAMURA, K. and TSUBOI, J. (1990), "The Report on Teleworking", in *Proceedings of the Pacific Telecommunications Conference "90,* Hawaii, pp. 716-721.

NATIONAL RESEARCH COUNCIL (1985), *Office Workstations in the Home,* National Academy Press, Washington, D.C.

NEWSLETTER ON COMMUNITY TELESERVICE CENTRES (1990), *Newsletter on Community Teleservice Centres,* Vol. 1, and Vol. 2, in press, Geneva and Odense.

NILLES, J.M., CARLSON, F.R., GRAY, P. and HANNEMAN, G.J. (1976), *The Telecommunications-Transportation Tradeoff,* Wiley.

NILLES, J.M. (1988), "Traffic Reduction by Telecommuting: A Status Review and Selected Bibliography", in *Transportation Research,* A, Vol. 22A, N° 4, pp. 301-317.

NORA, S. and MINC, A. (1978), *L'informatisation de la société,* La documentation française, Paris. English translation, 1980, *The Computerization of Society,* MIT Press, Cambridge, Massachusetts.

NORWOOD, H.K. (1989), *Principprogrammer* (The Danish Union of Clerical Workers' 1989 Declaration of Principles), Copenhagen.

OLSON, M.H. (1982), "New Information Technology and Organizational Culture", in *MIS Quarterly Special Issue,* pp. 71-92.

OLSON, M.H. (1988), "Organizational Barriers to Telework", in *Telework: Present Situation and Future Development of a New Form of Work Organization* (Korte, W.B., Robinson, S. and Steinle, W.J. eds.), North-Holland, Amsterdam, pp. 77-100.

POSTMAN, N. (1982), *The Disappearance of Childhood.*

QVORTRUP, L. (1984), *The Social Significance of Telematics,* John Benjamins Publishing Company, Amsterdam.

QVORTRUP, L. (1989a), "The Nordic Telecottages – Community Teleservice Centres for Rural Regions", in *Telecommunications Policy,* March 1989, pp. 59-68.

QVORTRUP, L. (1989b), "Community Teleservice Centres and the Future of Rural Society", CPF Conference, Swansea *Review of Current Experiences & Prospects for Teleworking,* (1990), a report prepared for Direction F, DG XIII, Commission of the European Communities, British Telecom and Business in the Community by Systems Synthesis Ltd. and Protocol Communication Ltd., Brussels.

RIJN, F.V. and WILLIAMS, R. (eds.), (1988), *Concerning Home Telematics,* North-Holland, Amsterdam.

ROBINS, K. and HEPWORTH, M. (1988), "Home Interactive Telematics and the Urbanisation Process", in *Concerning Home Telematics* (Rijn, F.V. and Williams, R. eds.), North-Holland, Amsterdam, pp. 185-198.

SCHWOHNKE, A. and WICKE, H.-G. (1986), *Teleheimarbeit als neue Rationalisierungsstrategie,* Pahl-Rugenstein Verlag, Köln.

SHIRLEY, V.S. (1985), "F International: Twenty Years" Experience in Homeworking', in *Office Workstations in the Home,* (eds. National Research Council), National Academy Press, Washington, D.C., pp. 51-56.

SHIRLEY, V.S. (1988), "Telework in the UK", in *Telework: Present Situation and Future Development of a New Form of Work Organization* (Korte, W.B., Robinson, S. and Steinle, W.J. eds.), North-Holland, Amsterdam, pp. 23-32.

STATE OF HAWAII, DEPARTMENT OF TRANSPORTATION (1990), *Hawaii Telework Center Demonstration Project.*

STEINLE, W.J. (1988), "Telework: Opening Remarks on an Open Debate", in *Telework: Present Situation and Future Development of a New Form of Work Organization* (Korte, W.B., Robinson, S. and Steinle, W.J. eds.), North-Holland, Amsterdam.

STERN, E. and HOLTI, R. (1986), *Distance Working Study,* FAST Occasional Papers N° 92, Brussels.

TOFFLER, A. (1980), *The Third Wave,* William Morrow and Company, New York.

VEDEL, G. (1987), *Danske kvinders distancearbejde,* Samfunds-litteratur, Copenhagen.

WAGEL, W.H. (1988), "Telecommuting Arrives in the Public Sector", in *Personnel,* Vol. 65, N° 10, pp. 14-17.

WIENER, N. (1950), *The Human Use of Human Beings. Cybernetics and Society,* Houghton Mifflin Company, Boston.

WEIJERS, T. and WEIJERS, S. (1988), "To Introduce or Not to Introduce Telework...", in *Concerning Home Telematics* (Rijn, F.V. and Williams, R. eds.), North-Holland, Amsterdam, pp. 413-422.

YAP, C.S., RAMAN, K.S. and LIM, G.K. (1990), "Alternative Work Arrangements for Computer Professionals", in *Proceedings of the International Conference on Information Technology,* Kuala Lumpur, pp. D.1-D.20.

ZIMMERMAN, J. (1986), *Once Upon the Future,* Pandora Press, London.

PUBLIC INFORMATION SYSTEMS
AND NEIGHBOURHOOD INTEGRATION

Bengt SÖDERSTRÖM
Director, Närbo
SWEDEN

Bengt Söderström, architect, works as a consultant in Närbo AB and is its director. Founded in 1986, it is a consulting company specialising in developing good neighbourhoods in housing areas, through developing public services and housing management. Närbo works mainly for municipalities and housing organisations, elaborating local programmes carrying out educational programmes. In co-operation with the Swedish Association for Local Authorities, Narbö is at present building up a structure for development and training in area-based working methods in social assistance, education and culture for the benefit of young people and the elderly.

INTRODUCTION

Strong faith in the potential of technology to meet varied needs and solve different problems is widespread.

Many predictions tend to overlook the importance of direct human contact and the strong restrictions inherent in traditions, administrative systems and other structural elements. This is particularly evident in the approaches being adopted to bring about stronger community feeling, local identity and co-ordination at local level.

The key words of my title – public information systems and neighbourhood integration – are not particularly clearly defined concepts. Public informations systems can vary widely, from municipal information systems aimed directly at keeping the public informed of local services, to general information systems such as tele-networks and cable TV, and commercial applications to provide better access to goods or services.

The concept of neighbourhood integration includes a desire to co-ordinate information to improve the finances and efficiency of various systems, and also to help the local population by providing access to many different information sources at one and the same place. The concept also reflects a criticism of the sectorisation and specialisation prevailing in society in general, and perhaps especially in the public sector.

This makes it difficult for citizens to obtain the service they need and to which they are entitled. But it also means that the interfaces between people are reduced and that our common responsibility for our surroundings is diluted. One result of sectorisation, specialisation and professionalisation is therefore that intended efficiency gains often give rise to many negative social effects. Another important aspect of neighbourhood integration is therefore an effort to revitalise local social contexts, to give people better control over their living conditions and to encourage shared responsibility for common, local affairs.

Three dimensions of a neighbourhood

The most accurate definition of a neighbourhood gives the neighbourhood a *social dimension*. According to this definition a neighbourhood is a place where you feel at home, feel secure and have an identity – where you are someone to the people around you. It follows that the neighbourhood is a small-scale unit, defined by informal links between people who live close to each other. In the neighbourhood we look upon ourselves as participants, acting out our daily lives privately and together. In this sense, a neighbourhood cannot contain more than a few hundred homes.

The neighbourhood can also be defined in terms of a *practical dimension*. It is the place where we carry out our daily work and where we receive public and commercial services. In this sense the neighbourhood is defined as an economic unit that has to contain a certain number of consumers. Normally, when defined in this way, the neighbourhood is much bigger and often contains several thousand homes.

Sometimes a neighbourhood is defined in terms of a *political dimension*. The neighbourhood is thus described as a place where people share a common interest in their surroundings, where voluntary work, community actions and public participation occur, regardless of ideological or organisational allegiances. In this sense, the size of a neighbourhood can vary considerably, from a group of houses to a township of several thousand.

Social needs in focus

These different definitions overlap. It is however important to bear in mind that first and foremost the neighbourhood is a social unit that should be able to fulfil basic human needs. The functions of a neighbourhood might well be described by using Maslow's hierarchy of human needs. It must give access to basic necessities and security, but it should also fulfil our need of trust, a sense of belonging, appreciation, dignity and self-realisation.

It is a sad fact that local action in a great number of suburban neighbourhoods increasingly has to aim at fulfilling basic necessities and the need to feel safe and secure. *Economic and social breakdown in peripheral areas of cities* is a problem that will have to be tackled in urban policies of the near future.

Demographic and social changes have also given rise to a *growing need for services* to care for the elderly, children, the ill and the handicapped. In all developed countries there is a common trend to reduce institutional care and to replace it with a dispersed network of social, medical and psychiatric services close to people's homes. For both economic and social reasons the ambition is also to integrate these services and to strengthen the links between professional staff and inhabitants of the neighbourhood.

There is also a growing problem of *loneliness* in modern urban areas. The city of Stockholm, where 51 per cent of households contain one person, tops the loneliness league ahead of Manhattan, New York. Meeting-places and different means of supporting social networks are becoming crucial elements in the urban fabric.

Social issues are becoming more and more important in urban policies. In an integrating world with growing competition between cities, *measures to promote economic growth* must be combined with ever more conscious measures to *distribute wealth* within cities. Information technologies can be used for both these purposes if they form part of a coherent strategy.

Different cultures

Both public and private services are provided by organisations which are formal and hierarchical and which generally see the inhabitants as consumers of services that have to be standardised in form and price.

In the informal neighbourhood culture, on the other hand, inhabitants look upon themselves mainly as participants in a local network of contacts and relationships between people with individual needs, interests and abilities.

The local culture is also characterised by attitudes which differ a great deal from those prevailing in the world of technical development. It tends to be a mainly female culture, with an emphasis on small-scale human networks, non-complicated practical solutions, a fairly short-term perspective and in general a very realistic attitude to economic and social facts.

These two worlds – the formal, professional organisation and the informal, local network – can interact in a constructive way, but this demands profound changes in attitudes and working methods within professional organisations. An ambition to do things *for* people has to be replaced by an ambition to do things *with* people, to become *enablers* instead of *doers*. Successful use of new technology on the neighbourhood level must take into account the special features of local cultures.

State of the art

Development of information and communications technology is very rapid. Fundamentally different systems and concepts are being developed side by side. Despite a strong desire to realise the advantages of co-operation between different system builders, in practice different organisations often build their own systems. This results in a very disparate picture and makes it difficult even to identify the main direction developments may take in the future.

This section presents some examples of information systems in operation or under development which may be of importance to services offered to citizens or which may affect living conditions in some other way at neighbourhood level.

Co-ordinated delivery systems

The best example of co-ordinated systems for the provision of local services is the increasingly sophisticated co-operation that is taking place in most OECD countries between post offices, banks and insurance companies. There is a rapid development towards companies able to provide a range of *financial and postal services* via a network of local offices. These services will gradually become available directly in homes via television and telecommunications systems.

In *France,* the Minitel now offers special services under the heading VIDEO-POSTE, giving residents the opportunity to receive information at home about different financial services, to make payments and transfer money, to buy travellers' cheques, to get information about the stockmarket and to communicate with the Post Office.

In *Sweden,* the National Telecommunications Agency has provided videotex since 1982. At present the service has 26 000 subscribers and is almost exclusively used by companies, but some small trials have been carried out in the private market. A new system called *TeleGuide* is just about to be launched in Sweden.

TeleGuide is a videotex service to enable householders with terminals and a special smart card, which holds a personal code, to perform bank transactions, consult telephone directories, order travel tickets, and order goods from mail order companies, from their homes at any time of the day or night. News and games will also be available. The TeleGuide service defines its business idea as ''to provide Swedish households with services they regard as meaningful or which simplify daily life in some other way''.

It is planned to distribute the first 50 000 TeleGuide terminals along with 19 000 devices to enable PC owners to access TeleGuide early in 1991. The coverage target is 500 000 households by the end of the 1990s. An annual growth rate of 50 000 to 100 000 terminals is envisaged. To achieve this the project plans to target existing customers of companies whose services will become available on TeleGuide. This may be compared with, for example, the strategy of full geographical coverage which the French public telephone company has for its Minitel terminals. An average total monthly charge of some SKr 150 is expected. This will consist of a flat-rate fee and time charges for use of the service, graded in bands.

Services directed to people's homes are of special interest to *help elderly or disabled people.* In many OECD countries, communication systems to call for help in an emergency are in operation. The concept of the ''interactive home'' could also be used to combat isolation. The Disability Information Services of *Canada* is an example of a telecommunications-based network service which facilitates the exchange of mail, has specialised electronic information/bulletin boards and allows electronic phone conversations. The Window on Technology Centre Project, initiated by the Ontario Ministry of Community and Social Services, uses a variety of ICT technologies to automate everyday living practices through voice commands and touch control.

ICT is also used to improve *transportation services* in local areas. Dial-a-ride systems exist in many cities in OECD countries. Examples are the *Christobald* system in the south of France and the *RUFBUS* in Hanover in Germany.

These transportation systems may also be linked to the provision of other public services. One example is *''Home deliveries''* – an ordering and distributing system for pharmacy and convenience goods in the city of Jönköping, Sweden, described below.

In co-operation with the County Council and the Jönköping municipality, the National Corporation of Swedish Pharmacies has since 1987 been involved in trials to develop the distribution of goods to patients being cared for at home. The Jönköping project now covers two care districts with some 17 000 residents.

It consists of a computerised order system for goods and transport services alongside a local transport organisation with a regular delivery schedule. The system guarantees high reliability by linking transport orders to goods orders.

At present there are 150 registered users in the system. About 520 orders and 165 messages are registered per month. Two pharmacies, two district nurse offices, four shops, the county council's central store and the transport service are linked to the system. The municipal home help service has five terminals. Altogether there are 16 system terminals and two links to other systems. Both PCs and minicomputer terminals act as terminals. All are connected via modem and open phone lines to a minicomputer located in one pharmacy.

The project concludes that communication should be based on local systems which can be linked via common communication computers. The home delivery service could also improve the use and profitability of existing local computer capacity such as shop computers which could supply information direct to the system. However, it is vital to develop a strong policy to ensure that individuals' privacy is respected by a home delivery system of this kind.

The use of ICT in medical care makes it possible to link local medical teams to hospital staff and to keep patient-held medical records. Examples are the North American decision-support nursing systems COMMES and HELP, designed to help local nurses and doctors decide on action in individual cases. Systems like these will make it possible to create *decentralised and flexible medical and health care organisations*. A rapid development in this direction is taking place in many OECD countries.

Services directly to people's homes do not necessarily have a favourable impact on the neighbourhood level. Although services may be provided more quickly and cheaply, they may also bring about poorer common local services, fewer places to meet and more isolation. Other solutions, in which services are provided at local service centres, are of great interest.

There are several focal points at the neighbourhood level that can serve as *integrated service centres*. Local shops, tobacconists and cafés are obvious examples as are post offices. Retail trade and other services can use ICT to obtain at the local level economies of scope instead of economies of scale. There are several examples in OECD countries of co-operation between commercial organisations at the local level.

One example is the distribution of *theatre tickets* through post offices in Sweden. To maintain a wide network of local offices the postal service is looking for ways of offering customers additional services. In 1978, the Swedish Post Office started discussions with the Stockholm Opera, the Stockholm Concert Hall, Sandrews (representing two private theatres) and Skådebanan (a sales organisation for theatres) on the joint development of a common ticket booking system based on modern computer technology.

At present, 700 post offices are connected to the booking system created (Ophelia) and more are being added gradually; by 1992, about 900 will be connected. The number of theatres and other venues connected is 17 in 1990. The system is also used for booking and ticket distribution for special events. The average booking volume is currently 2 000 tickets per day, or 30 per cent of the total capacity of the venues connected.

The heart of the system is a large data-processing installation at the Post Office in direct communication with the theatres connected to the system. The theatres each have a local computer, display screens and a ticket printer. The local computers can communicate directly with the central processor, which in turn is in communication with the post offices connected to the system.

Software is supplied by the Post Office, which also assists in adjusting and starting up the equipment and routines. The Post Office also provides training courses for users of Ophelia; computer equipment is supplied in co-operation with Honeywell Bull.

The Post Office gives marketing support to the theatres at its outlets. A survey shows that theatre-goers greatly appreciate being able to buy tickets where they live the same day they book. This also benefits theatres since it increases their cash flow and makes them more certain that bookings really will lead to sales.

Totalisator *betting on horse racing* is another example of a local service that is often distributed in co-operation between different organisations. There is a need for an extensive information network to provide adequate services for horserace betting.

In France, betting is done at the local café. A betting system uses the France Telecom system to provide 6 500 terminals in cafés for the registration and payment of bets.

In Sweden, betting is performed at the tobacconist's shop. It is administered by ATG, a company jointly owned by the national federations for flat racing and trotting. There are some 1 050 race meetings per year, three per day, and ATG's turnover in 1989 was SKr 5.9 billion. The public can take part in off-course, pre-race betting through any of ATG's local outlets. Some 1 100 of these have terminals connected to the AGENT system via 140 of the National Communication Agency's DATEX lines. The system came into operation in 1987, and is planned to include 2 200 terminals covering the whole country. During peak periods the system handles 40 transactions per second; however, in normal operation only 50 per cent of computer capacity is required.

This form of betting involves a great need for up-to-date information on horses, current odds and results. This information is made available to the public on several presentation systems.

RADTEX at most ATG outlets presents meetings, odds and results on two TV monitors, using information distributed via the network for national TV broadcasts using the same technology as TEXT TV.

Video programmes are sent on the TV network once a week showing the past week's race meetings. The transmissions are coded so that they can be taped by ATG outlets.

VIDEOTEX supplies information on horses and betting. Using a modem, customers who connect a PC to their TV can access information in ATG's database on bets and race meetings.

Local information centres

In many OECD countries various kinds of local information centres are set up in order to give the public better access to municipal services. Most of these are in fact not based on a neighbourhood strategy, but nevertheless they illustrate possibilities of integration and decentralisation that can be further developed. Some examples are presented below.

Service shops in Denmark

At least 15 Danish municipalities are experimenting with different forms of service shops, located either centrally or in suburbs.

In one municipality, politicians have demanded that 20 municipal advisers should be trained to provide service from several departments. The municipality in question is developing generalist training and is convinced that future leaders of the public sector will be found among generalists with broad knowledge and the ability to place the customer in the centre.

In June 1989, Skaelskoer opened a service shop staffed by five people recruited from the tax department, the finance office, and the social services and school departments. Services include applications for pensions, subsidised travel services and day-care places; registration of name changes; social insurance certificates, *etc.*

The service shop in Roedovre, in operation since October 1988, can deal with all simple tax, civil registration and health insurance cases and its six employees act as citizen advocates, also channelling citizens to many other municipal services.

One service office in Ravnsborg, linked to the Town Hall by computer, serves an island with a population of 700, dealing with 90 per cent of the islanders' municipal contacts.

Citizen offices – Bürgeramt – in Germany

The citizens' office, Bürgeramt, set up in 1984 by the city of Unna, which has a population of 58 000, illustrates the German experience.

Unna had started a project called Open Town Hall in 1978 to orient the administration towards citizens. The present citizens' office consists of a central office in the Town Hall and three suburban offices with some 20 employees in all working in teams of three to four, with complementary skills.

The three goals of these offices are:

– to try to reduce bureaucratic time wasting and give the public better service;
– to determine how computers best fit into local public administration; it was clear from the beginning that computers were essential for the citizens' offices; and
– most important of all, to introduce changes with the help and participation of the people affected, the citizens and front-line staff.

This led to an ambitious programme of consultation and surveys.

The duties of the offices have expanded gradually from the initial package of housing registration, issuing passports and identity documents, issuing permits for foreign citizens, handling postal voting *etc.,* and supplying information on cultural and sporting events. The offices took over simple social services cases, order and security, and subsequently more complicated work like issuing commercial licences and applications for housing allowances.

Neighbourhood offices in United Kingdom

Neighbourhood offices have been in operation in some English district councils for some time. They are represented here by the case of Walsall, near Birmingham, with a population of some 250 000, which set up 31 neighbourhood offices in 1981. A total of 300 people work in the offices.

The political goals were to move local services closer to people so as to create more interest and participation among citizens. Service goals included better physical accessibility, a more personal and informal service, cross-sector service, greater status for the work of front-line staff, and the freeing of administrators to work more with customer contacts; staff at neighbourhood offices were to be generalists working in special teams.

So far the offices have concentrated on housing services previously run by three departments, including rent payments, housing benefits, repair requests, housing allocation, applications for repair loans, applications for home help, meal deliveries, aid to the disabled as well as advice on social benefits. The offices have advanced computer assistance with a calculation programme for housing benefit and an expert system for social benefits, for use with customers, in addition to traditional case-handling systems.

The CISP project in Sweden

The CISP project – in the muncipality of Härnösand in the northern part of Sweden – aims to develop computer use in contacts between front-line staff and citizens in order to support the development of customer-oriented organisations in the public sector. Public handling of information must therefore shift from being administrative and producer-oriented to being strategic and customer-oriented. The main concern of CISP is how computers can be used to develop the interface between the customer and the public employee, for example in the dialogue that takes place between them. Participants in CISP are the Regional Social Insurance Office and Labour Board; the State County Administration and the County Council; Härnösand Municipality; the National Telecommunications Agency; The Swedish Association of Local Authorities and Kommundata, the authorities' company for computer services; and the National Agency for Administrative Development.

The project began in 1986 and, to date, has resulted in the development of basic information systems for certain groups of customers and in increased knowledge about customers' needs and about implications for administration, work organisation and staff training *etc.* As part of the project, a pilot scheme was started in Härnösand in August 1990. Work is also in progress with a view to introducing the system at other locations.

To be of interest to citizens a Citizen Information System must provide useful and specific information and knowledge. The system should be able to answer questions like:
- whom should I contact?
- what kind of service does the authority provide?
- how do I start?
- how can I influence the authority?
- what kind of service is available in my situation?
- how is my particular case dealt with?

In the initial phase, work has focused on building a simple information system for customer groups like pensioners, parents, young people/unemployed, and home-builders. In several areas work has already gone beyond simple reference and public information. There are already computation systems which can give individual advice as well as support in completing forms so that the citizen can take up a matter with the proper authority and have all of the necessary material on hand.

For a long time to come, most public services will be handled via an official, who can also support computer-based information with personal knowledge. The CISP system is not for direct use by the citizen; instead the point is that staff groups with customer contact such as receptionists, telephone operators and front-line staff should, with the support of the technical system, be able to give the customer a better service – in breadth and depth – than today. The official uses the screen to present information to the customer, and they look at it together. Naturally a totally different interface will be needed in future, including photographs, graphics and sound. The interface may also be customised; if groups of customers have difficulty reading text the presentation may rely more on pictures and graphics.

The CISP project is developing systems on computers using the Hyper-Card programme. The system is so user-friendly that customers at public terminals could easily use it themselves to obtain simple information not requiring knowledge in depth about rules and regulations. The system allows feedback from the customer to influence the service. For this purpose a subsystem has been built, containing a brief evaluation form which the customer can complete directly on the screen. Another possibility is for customers to enter their own comments which can be stored or sent to the official concerned. Alternatively the official can be called via a button on the screen.

The neighbourhood service project (NUDU) in Umeå, Sweden

The NUDU Project was started in autumn 1986 and was terminated at the end of 1989; ongoing activities were either incorporated in more permanent organisational forms or discontinued.

The project has generated knowledge about the opportunities for co-operation, practical applications of new technology, and the role of information technology in society and what is required to make sure its positive potential is realised. The project comprised:

- a *neighbourhood centre* – a local communication centre for housing services, use of a computer-based message system and fax for communication between the neighbourhood centre, the health centre and housing management, *etc.;*
- a *same-day reply service* – reception for residents, use of an information database, computer-based message system and fax for more effective service to the public – questions and answers;
- *local TV* information to households via the cable network;
- *mobile/cellular telephone in home help services;*
- *messenger car services* by co-operation between home help services and housing management.

The project was managed by a management group with representatives of Umeå municipality, the local housing trust, the Västerbotten region of the National Telecommunications Agency, and organisations.

The district of Marieberg (population 8 000) was considered to be a natural location for the centre. The service level in the area is poor, there is a need for a meeting point, walking distances are long, local information is poor, there is a need for better co-operation in public services, *etc.* The sheltered housing block seemed to be the appropriate location; it had plenty of space, a bold, development-minded manager, and a manned reception which could form the basis of an information centre.

To prepare the project, open study circles were started in autumn 1986, and everyone in the area was invited to attend – pensioners, people active in local organisations, home help service staff, ordinary residents *etc.* The subject of the circles, which were led by university teachers without any links to the rest of the project, was information technology in housing.

The thinking behind the circles was to create a point in the residential area which people would visit to meet one another, eat, obtain information and make friends. No specific view on how information technology might be used to this end was advanced. However, the circles did express the fairly strong view that it was not enough to contact a neighbourhood information centre for local information.

People had TVs at home and wanted to use them. In the original plan there was no thought of local programmes on cable TV. This came about as a result of the discussion in the circles.

At the National Housing Exhibition in Umeå in August 1987, a housing services information centre was set up, based on information technology. The development of this prototype provided important lessons and experiences about the technical, administrative, human and political problems to be solved. In autumn 1987 and spring 1988, work continued on building up, developing and manning the information centre. Much effort was devoted to spreading the word about the centre, with the help of the new cable TV system, which could reach 300 households.

Gradually, the information centre became an integral part of the housing block, and the staff and residents became more positive and interested. A large proportion of the residents (50-60 per cent) watched local TV, and spontaneous visits from residents in the area began to increase. Local TV coverage was received very favourably and helped to keep the staff of the neighbourhood centre informed. But it required a great deal of involvement by the service staff. To ease the burden it was hived off from the local centre.

The equipment which remained in the neighbourhood centre consisted of a PC with a modem and fax, *i.e.* equipment mainly used for communication. It was increasingly realised that any day nursery or library could act as a neighbourhood centre, with the aid of fairly simple equipment, as long as neighbourhood TV was extended to the whole municipality.

Trials with mobile cellular phones were also started in 1987, after the transmitting station had been built and the newly developed telephones acquired. From the very beginning reactions from participants were positive.

The success of neighbourhood TV and cellular telephones helped to strengthen confidence and determination over what appeared to be the most difficult part of the project – providing municipal and housing company services in a good and flexible way.

Some conclusions of general interest can be drawn from this project.

Municipalities, county councils and housing companies often have completely different centralised computer environments incapable of communication with each another. Much time and effort is required to set up the technical means to communicate between them.

The new communication methods have been of great help in developing contacts between the neighbourhood centre, housing and messenger services, pharmacists, district nurse and municipal departments, especially social services. Most of the information has been about individuals' needs, and one result is that the district nurse has opened a clinic at the housing block where the service is based for a few hours a week.

The introduction of new means of communication is indispensable in promoting practical co-operation. However, it requires development of a technical network and new patterns of behaviour, which take some time and pose problems. It is necessary to:

– apply recognised technical standards and deal with strongly territorial organisations and supplier monopolies;
– ensure strong and consistent messages from management in the organisations involved;
– use a coherent method and teaching approach.

It is hard to get people to visit a new place they are not used to. An introduction really ought to be built into the natural contacts they already have with the municipality and the housing company, as it is with day nurseries and libraries.

The public asks for straight factual information; only a few people have well thought-out questions. Often a consultation is needed to find out what they really want; at best this can lead to quick help either based on the centre's own knowledge or through contacts with the right person. It is probably impossible to tailor and maintain databases and routines to answer the majority of the everyday questions people have. One popular service is the housing allocations list of vacant flats.

After local cable TV had been broadcasting for six months, 60 to 70 per cent of residents were viewers. When the area was connected to the Telecommunications Agency's cable network in 1988 the number of viewers increased rapidly to 1 200 and information started to flow in from local associations and municipal departments. There were favourable reactions from the public, and local TV now runs a 15-minute programme on events in Umeå, useful local information and what's happening in three housing areas.

The project has given positive results, mainly in the form of improved neighbourhood services such as more accessible and flexible home help staff, simpler and quicker contacts with municipal departments and health centres, better use of resources, improved knowledge about the area and a greater sense of being at home there.

Other examples

There is a wide range of ways of creating local information centres. In France (Brest), a register of urban services has been developed on teletext, enabling the general public to have easy access to all aspects of everyday dealings with the administration (employment, education, housing, health *etc.*). In Strasbourg, a system named ALSAMATIQUE has been installed, giving the public wide opportunities to ask for information on municipal services, entertainments, *etc.* and also to receive personal messages. The system is based on the Minitel communications system.

In France and in some other countries, municipal information is given in the form of "electronic newspapers". Such displays are situated in urban areas where pollution is heavy, and used specifically to give the public up-to-date information on the current environmental situation.

Libraries, schools and universities provide good opportunities to connect local information centres to multi-purpose networks of communication. The creation of distance learning and training facilities, such as the "Learning Centre" in Ottawa, Canada and the "Intelligent Plazas" in Kawasaki, Japan, gives us a hint of future possibilities on the neighbourhood level.

Local TV

TV is becoming more internationalised as a result of the increasing range of satellite channels. However, the expansion of cable TV is also creating new opportunities for local TV programmes to strengthen the sense of community at neighbourhood level and to spread local information. Sweden has had considerable success in setting up local TV stations offering access to community and non-profit interests.

The cable TV network in Sweden has grown rapidly in recent years. In 1990, some 1.15 million homes are connected to one of the existing cable networks. The largest, with 70 per cent of households connected, is owned by the National Telecommunications Agency; private cable companies have just over 20 per cent and public housing companies 8 per cent.

Cable networks give households access to locally produced material in addition to national and international channels. There are special regulations for local programmes to more than 100 homes, including the need to set up a local cable broadcasting body, which has to be representative of the community, open to new members and prepared to give access to representatives of various local interests and opinions.

Local interests should be able to take part in programme making and the most varied opinions should be expressed. This means participants with limited resources and experience of the medium should be encouraged. Both regular and occasional programme producers should be able to take part.

The government gives some financial support to activities that lead to locally produced programmes being broadcast on the cable network. The National Cable Television Board sees substantial scope for ideas and proposals to improve local TV production and help local cable channels to start up in many places.

In September 1989, there were 17 local cable bodies with permits, and new bodies are planned in some places. Ownership varies: for example, in Stockholm local associations and organisations dominate, whereas in Malmö and Göteborg companies participate alongside local associations. Housing companies, local authorities or local higher education institutions can also play a major role.

The coverage of local channels varies with the proportion of households cabled. In Botkyrka, a Stockholm suburb, 80 per cent of households are cabled, while the figures for medium-sized towns like Norrköping and Linköping are 18 and 15 per cent. Usually distribution is on the National Telecommunications Agency's network. It is expensive to distribute programmes from a studio to small, distant cable networks in a municipality.

Housing companies exercise a decisive influence at present, often initiating cable development in order to create a sense of community and make their own housing areas more attractive. The tenants' associations and HSB housing co-operatives often join for the same reason. Business involvement is increasing and interests behind new stations usually include local companies or chambers of commerce.

Production conditions vary, but only a small part of the costs is reflected in the budget, owing to a mixture of voluntary and paid work, borrowed premises and equipment. Transmission costs depend on how the local body's finances are organised, the size of membership fees, income from text information and the number of employees. At present costs are highest in Göteborg and Malmö.

Membership charges, when levied, vary from a nominal SKr 100 to SKr 60 000 (in Västerås). There are instances of grants and favourable loans between members of local cable TV associations, but funding support from local government is rare. Municipalities generally participate on the same conditions as other interests, although in Landscrona cable TV is part of the municipal culture department. Local government information is common on cable TV, as is free support from municipal audio-visual centres and culture departments. Local political debates are less usual.

Participation of local political parties and of trade unions varies. In several places public agencies and institutions such as police and labour exchanges have their own information programmes. Film and culture associations play a relatively small role, while sport is more prominent, and local universities are often represented. Co-operation with adult education exists in a variety of forms, while co-operation with schools and childcare centres is rare. Perhaps a change can be expected here.

Telework

Telecommunications and computer technology offer new potential to realise the idea of local workplaces by making long-range communication between workplaces possible. At a neighbourhood centre – using advanced computer equipment – people can work for a variety of companies at a distance from their main offices. This type of workplace can be located in or near the home town or housing district of the staff concerned.

Telework is dealt with in another section of this publication. Just a few experiences from an attempt to create a neighbourhood working centre in Sweden will suffice here.

The project organisers chose Nykvarn, a community of 6 000 in the south of greater Stockholm, as the location for the centre. It was planned in 1981-82 and opened in October 1982 with nine workplaces which were used by ten people including a researcher from the Nordic Urban and Regional Planning Research Institute. In addition, two workers from a job creation programme helped to provide support at the centre. In April 1984, Nordplan broke off their involvement a few months earlier than planned for financial and practical reasons, but the centre continued in operation for another 14 months.

Access to computer technology was necessary for most workplaces located in the Nykvarn centre. Communications with head office were important for seven out of ten, and one of these broke off the experiment after a short time since presence at the head office was felt to be important. Some had problems with physical transport of materials, long reply times and other technical defects. Closeness to other tasks and equipment in the centre stimulated social contact and spread knowledge. For some it resulted in disturbance.

The centre had to fulfil many conditions – such as individual flexibility at work, access to new technology, the opportunity to test a new business idea, and the right environment for a new company. This led to very varied expectations, and in part to the low attendance, 30-45 per cent of full-timers. Low utilisation meant the centre could not always provide the expected social atmosphere and colleagues to discuss and solve problems with.

Some people continued to spend most of their working day at their principal place of employment; others were given tasks of such a different nature that the centre became their principal workplace. So the original idea – working at a distance from the principal place of employment – was scarcely tested at all in reality.

The experiment shows that some groups and individuals who can make use of the independence which comes from working away from the principal place of employment take great interest in developing the idea of neighbourhood working centres. The centres are also a solution for companies which, for marketing or recruitment reasons, want to expand their activities without too heavy an investment. Neighbourhood working centres could provide additional working places in small areas.

On the other hand, there is probably no benefit to be gained by separating out routine jobs and allocating them to neighbourhood working centres, thereby worsening the labour market for weaker groups such as people with limited training, women, part-time workers, *etc.*

The cost of extra premises, supervision of work and the disadvantages of rigid organisation probably outweigh the savings to be made by lower wage rates or lower rents in outlying areas.

Assessment

Conflicting trends

Public efforts to strengthen the neighbourhoods are counteracted by other trends in society. In many OECD countries, public services have to be cut for economic reasons. Commercial services are undergoing continuing development towards concentration in larger service centres, easy to reach by car. Large amounts of money are spent on publicity and advertising promoting lifestyles oriented towards personal professional success and private consumption.

To meet the problems caused by these conflicting trends a redefinition of the tasks of different producers of services is taking place. Public services are being integrated and decentralised into smaller units with some degree of self-management. Public service organisations are being privatised or exposed to competition from private enterprise. Public and private housing organisations are expanding their areas of action.

Most service providers are developing systems that will make it possible to order and deliver services directly from and to the individual home.

Integration of existing resources

A policy to make use of new information technologies must take into account the conflicting trends in current development in the urban fabric, in the network of services and in the roles of different organisations. A strategy to strengthen neighbourhood integration has to be based on good knowledge about real local needs, as they are defined by the inhabitants themselves, but also on an understanding of the way in which the local informal society works.

In many OECD countries, projects to strengthen local neighbourhoods have been carried out. In some of these, integrated public information systems are an important element. Very often, the ambition has been to create a community in which the residents can easily get in contact with all kinds of public services and get the information and assistance they need in one place.

A common experience seems to be that demand for this kind of service is less than estimated. Probably the reason for this is that very often these new services do not really integrate existing services. They are additional services on top of all the others, but they are not the result of a more coherent approach to the needs of the inhabitants, and only in a few cases have they made a real impact on public services.

If it is possible to create co-operation between education, social services to children and the elderly, culture, leisure activities and housing management, then it is also possible to create on the very local level a supporting organisation, based on personal contacts and an open and integrated use of premises.

Here the local perspective is important. Today, in a Swedish housing area with 1 000 apartments, some 200 people work in child day care, care of the elderly, schooling, recreation services, health and medical care, housing management and park maintenance. By working together more closely and with greater direct influence and participation by residents, better use can be made of resources. Such an approach also contributes to strengthening people's sense of identity, responsibility and involvement.

Services and management

While the commercial sector has been quick to apply new technology, the public sector is still only beginning to give citizens better service with the aid of technology.

In most developed countries, however, changes are now speeding up in the public sector. Here again there are conflicting trends. On the one hand, we can see efforts to rationalise and increase efficiency on traditional lines derived from industry. This work is largely directed towards better control and management of activities within the framework of the existing sectoral organisation. Important factors are setting up decentralised work units, defining goals more clearly, and better monitoring and control.

These efforts have a strong impact on the information systems built up in the public sector to steer activities. In this respect public bodies resemble businesses; the search for a careful adjustment to the needs of their own activity makes co-operation with other organisations more difficult. And the ultimate result of this trend is privatisation of public activities, as is happening increasingly in most industrialised countries.

A development of public services on the basis of a neighbourhood ideology makes different demands on information systems. First, information systems in various local authority departments must be co-ordinated so that it is possible to use common resources in a simple way while maintaining good control of factors such as finance and quality. Secondly, greater demands are made on internal communication between different staff groups, between residents and staff, and between residents concerning local events. These demands mean that new approaches are needed for information systems and that the work of the front-line staff has to be upgraded. Attempts to build up such locally oriented information systems have only just started in some countries.

Policy options and recommendations

Modern information technology creates new opportunities for co-ordination, integration and decentralisation. Technology, used in a purposeful way, provides opportunities to create more vital local communities, in which citizens have good access to information and services on their own terms. This is a context which gives citizens a much greater chance to decide on local affairs.

However, technology is only a tool. What will be decisive for future developments will be: the vision which will guide the development work of community planners and system builders, the way in which individuals and households will behave, and the success that will be achieved in attacking the institutional and administrative frameworks which impose restrictions on development work and co-operation.

There are strong reasons for building a more conscious neighbourhood policy in the developed industrialised countries. New structures must be created to meet people's need for community, security and identity. We must find new forms to help people find practical expression for concern, solidarity and common social responsibility. The neighbourhood is a natural base for such broader community among people.

In the modern welfare state, public institutions have access to substantial resources which can be used to support such a development. In an open, neighbourhood-oriented approach, working*with* residents rather than *for* them, child day care, schools, recreation services, health and medical services, *etc.* can help to strengthen contacts between people and to increase residents' participation in community life. But a conscious policy and major changes are also needed to create a context for citizens to become participants who accept responsibility instead of passive consumers of commercial and public sector goods and services.

Where there's a will technology can lend a helping hand. Some examples of important development areas are set out below.

Administrative systems for co-ordinated planning and management of local public services

A policy for neighbourhood integration demands that centralised, bureaucratic management systems be replaced by co-operating, dynamic organisations of a manageable size. Field staff must be given the opportunity to decide in direct contact with their ''customers'' how staff resources should be assigned, and how premises and available funds should be used. This requires new administrative systems capable of linking the integrated local perspective to the existing administrative systems and information systems for data on population and the built environment. Such local systems must be created in order to provide simple ways of testing the implications of alternative actions; and they must be illustrative and easily used by people with limited experience of finance and administration.

Information systems which promote a sense of community

Today we are being swamped by news about events all over the world. The development of TV and telecommunications will soon give us almost unlimited access to global information and communication round the clock. At the same time we can be completely unaware that our old neighbour is sitting alone and isolated in his apartment and we can completely lack contact with the environment and culture our young people live in.

This is a dangerous trend. One of the great challenges for modern mass communication technology is to find ways in which technology can help to strengthen human contacts in the neighbourhood community. The technical solutions already exist. The main thing now is to find appropriate administrative forms and economic resources to stimulate local initiatives to set up neighbourhood communication channels and use these unaided.

Experience from the Scandinavian countries, among others, shows that people have great interest in local affairs. However, we have to be aware that expectations of quality today are high, which means costs will be quite large. This then leads to a need for co-operation between professional and voluntary organisations which has to build on an openness about using one another's resources and competences. The public sector's activities can play a major role in promoting the construction of local communication systems.

Systems for co-ordinated orders and delivery of goods and services

Commercial services continue to be concentrated in large shopping centres in locations with good communications. Local services are becoming thinner and thinner. This presents difficulties to many people who are dependent on access to goods and services close at hand, elderly people for instance.

Public costs for help to elderly, disabled and chronically ill people are rising as local services thin out. Local shops also play an important role as meeting points and informal information centres – functions which are lost when the shop closes and is usually not replaced.

It is not inconceivable that local service centres could be set up, where both commercial and public organisations could offer their services close at hand.

It is not clear how far we can go down this road. Experience from many countries, however, shows that the public sector can make great gains in terms of both quality and costs. Evidently co-operation with certain commercial companies offers great advantages, including support for medical care in the home and other assistance to the sick and disabled.

However, this experience also points to difficulties which must be overcome. Activities which are exposed to competition find it difficult to show the desire for adaptation and openness which is needed to achieve co-operation. Activities in the public sector find it difficult to overcome bureaucratic obstacles and make front-line staff take greater interest in their work, which is necessary if local services are to be improved.

Successful development work directed towards co-ordinated local service systems probably requires a force sufficiently powerful to pull others in the same direction. Social services and home care are of a size and influence to be such a co-ordinating force. However, the internal co-ordination and conscious strategy which would be needed are still lacking. Yet there are prospects of interesting development work in this field.

Means to promote personal contacts

The most effective information system at neighbourhood level is probably the local café. If the goal of neighbourhood integration is to be achieved then technology must be developed in such a way that it brings people physically closer together and helps us to make contact with one another. Culture and leisure are therefore important parts of a strategy for developing local community spirit. New information technology can also be of use in this area. In several countries one interesting trend is that libraries, schools and adult education centres are becoming local cultural and communications centres.

Some lessons

Many of the development projects described here have started with a strong faith in the potential of technology and have ended with increased insight into the great importance of the individual. The focus must be on those who will benefit from development work. Too many development projects are based on the producer's wish to make his work simpler instead of on the customer's desire for better service.

Activities in the field cannot be altered without management support. Many attempts at local co-ordination have failed because the managements of the organisations concerned have not understood or accepted the idea.

Staff must share in development work. Good technology operated by people with insufficient motivation or competence is a disaster. Participation, support and training are needed. So is time.

Public activities must be accessible, visible and inviting. Information and marketing are important components in launching all new activities. Monitoring and ongoing evaluation must be natural elements.

Staff must be given the opportunity to adapt activities to customers' needs and wishes, and they must have channels for proposing more substantial change.

REFERENCES

The home deliveries project
Ingvar Mohlin
Apoteksbolaget (The National Corporation of Swedish Pharmacies)
S-105 14 Stockholm
Sweden

Theatre tickets in post offices
Johan Weine
Swedish Post Office
S-105 00 Stockholm
Sweden

Bookmaking for horse races
Claes Heijbel
ATG
S-161 89 Stockholm
Sweden

Citizen information system project (CISP)
Carl Öje Segerlund
Swedish Association of Local Authorities
Box 3014
S-871 03 Härnösand
Sweden

The neighbourhood-centre (NUDU) project
Ulf Brånell
Kommunkansliet
Umeå municipality
S-901 84 Umeå
Sweden

TeleGuide
Göran Tamm
TeleGuide Scandinavia
Box 1509
S-171 29 Solna
Sweden

Nykvarn telework project
Ulf Ranhagen
VBB
Box 5038
S-100 42 Stockholm
Sweden

Part 2

APPLICATION OF INFORMATION TECHNOLOGIES IN URBAN SERVICES

INFORMATION SERVICES
AND LOCAL ECONOMIC DEVELOPMENT

Mark HEPWORTH
Centre for Urban and Regional Development Studies
University of Newcastle upon Tyne
UNITED KINGDOM

Mark Hepworth graduated in Economics from the University of Warwick, United Kingdom. His doctoral thesis at the University of Toronto focused on the geography of the Canadian information economy. Currently, he is co-ordinator of the multidisciplinary Doctoral Programme on Information and Communication Technology at the University of Newcastle upon Tyne. In addition to the recent book "Geography of the Information Economy", Mr. Hepworth has published widely on the spatial impacts of information technology across the fields of geography and planning, economics and communications. His new research on transport-telecommunications interactions "Wheels and Wires", will be published as a book. He has recently established a London consultancy firm, CITI, which specialises in applied social, economic and technical research in the fields of road transport informatics, urban innovations, geographical information systems and the global information economy.

INTRODUCTION

During the last two decades, information and communication technologies (ICT) and information services have become increasingly central to economic growth and employment generation in all OECD countries (OECD, 1981, 1986). Research shows that the geography of the "information economy" within countries is dominated by large metropolitan cities and urban centres within regions (Castells, 1989; Hepworth, 1989a). In response to these economic and technological changes, local authorities in the urban areas of OECD countries are pursuing a wide range of ICT-based initiatives which involve the provision of information services to firms and other sectors. This paper reviews these new directions in local economic development initiatives in order to highlight their potential importance as policy options.

The advent of the post-industrial information economy, with its highly dynamic environment of technological change, market transformation and internationalisation, presents new and different challenges for local authority policy makers. In commenting on the challenge to town planners, for example, Howkins (1987:427) observes:

"The old-style planner talked about physical zoning, the balance of employment, housing and open space and traffic flows. The new-style planner... has to consider the configuration of electronic systems and Local Area Networks and the provision of bandwidth to each urban area. The town planner dealt with the stocks and flows of vehicles. Today's public authorities have to face the stocks and flows of information".

Planning for the information economy, through ICT innovations, is rapidly becoming an important area of policy development in local authorities. However, these policy activities are still new ground for authorities, and most ICT-based initiatives began to emerge in the 1980s, as the French national contribution to the OECD surveys tends to show (see, for further UK evidence: Hepworth, 1989b). The OECD surveys also strongly suggest that ICT-based initiatives will evolve as the cornerstone of local economic development strategies well into the future. A systematic consideration of current policy directions in this area is therefore both timely and relevant.

This paper concentrates mostly on the "services" component. The basic overall concern of this paper is the nature and substance of ICT policy initiatives which are explicitly aimed at promoting economic development in urban areas.

The range of current applications

The rise of the post-industrial information economy in OECD countries has generated a plethora of ICT-based policy initiatives at the local level. The first part of this section outlines some structural changes in the information economy which have provoked these different types of initiatives (*i.e.* as solutions to general problems confronting local authorities). Next, attention focuses on the nature and form of specific initiatives, and current policy directions are illustrated using the results of the OECD's special surveys and other sources. The last part of the discussion outlines the institutional framework of ICT-based initiatives, especially the role of local authorities in collaborative arrangements between the private and public sectors.

Structural changes in the information economy

Let us now look at some key aspects of structural change in the information economy, as a basis for understanding the current thrust of ICT policy initiatives in local government.

First, the last two decades have been marked by the rapid globalisation of the information economy, as a growing and significant part of international trade and foreign direct investment (Howells, 1988).

The major ICT innovations supporting these economic changes are private computer networks operated by multinational companies, and public computer networks operated by the telecommunications carriers (OECD/BRIE, 1989; Moss, 1987; Hepworth, 1989a). These globalisation trends have, of course, been accelerated by the Single European Market ("1992"), the "revolution" in world capital markets, the expansion of international trade in services, and new ways of organising multinational production.

The general implication of these global trends, across OECD countries, is that ICT-based policy initiatives to promote local economic development in the OECD Member countries are becoming more international in scope.

Second, and partly as a result of globalisation trends, firms in all sectors are increasingly confronted by higher levels of uncertainty in product and service markets. In OECD countries, competitive conditions in industries and markets are being radically altered by the deregulation and liberalisation policies of national governments, which are further reinforced in the European Community context by the transition to the Single European Market. As a result of these market uncertainties, firms are confronted by higher transaction costs (Williamson, 1985) – that is, the costs of searching for, negotiating with, and monitoring and enforcing transactions with other businesses (*e.g.* supply contracts), as well as with consumers (*e.g.* advertising costs), employees (*e.g.* salaries and working conditions) and governments (*e.g.* complying with national technical standards).

The basic resources that firms set against transaction costs are informational, including information capital (ICT and offices) and different categories of information labour (*e.g.* contract lawyers and sales clerks). The general implication of rising transaction costs in the economy is that ICT-based policies to promote local economic development are increasingly being directed towards "subsidising" the informational needs of firms.

Third, the new ICT provides opportunities for firms in all sectors to introduce product and process innovations, as a basis for greater competitiveness, efficiency and flexibility. In manufacturing industry, "just-in-time" methods based on inter-organisational computer networking (*e.g.* electronic data interchange) are supporting flexible specialisation between firms, whilst flexible manufacturing systems based on industrial local-area computer networks ("LANs") are enabling firms to obtain economies of scope (rather than scale) in variety-oriented, batch production (Piore and Sabel, 1984).

These same trends are evident in the service industries, where IT has transformed the "production line" in financial services (*e.g.* electronic funds transfer), transport (*e.g.* electronic route guidance), retail trades (*e.g.* point-of-sale networks), tourism (*e.g.* advanced ticketing systems) and other sectors. The general implication of this technological revolution in the economy is that ICT-based policy initiatives are increasingly directed towards improving the innovative capacity and information-intensity of the local economy.

Fourth, current patterns of occupational change in the labour force of OECD countries indicate that job growth is concentrated not only in "information occupations", but also in lower-order service occupations (Castells, 1989; Hepworth, 1989). These trends raise fundamental concerns over the future quality of employment opportunities and social inequalities, as well as standards of vocational education and manpower training (Schmandt and Wilson, 1990). Their general implication is that ICT initiatives are being directed towards increasing the social wage of the local population (better public consumer services – see innovations in other functional areas) and improving delivery systems in education and training.

Whilst the above structural changes are generating ICT-based policy responses in all local authority areas, the uneven spatial development of the information economy and the intensifying inter-urban competition for new investment and jobs have disproportionately increased the "stakes" or risks associated with economic development initiatives in peripheral regions of OECD countries (Harvey, 1989; Hepworth, 1990). For example, recent survey research on urban telecommunications policies shows that "peripheral" local authorities in northern England (*e.g.* Sheffield and Manchester) and Scotland (*e.g.* Glasgow and Edinburgh) are far more pro-active in terms of economic development-related ICT initiatives than "central" local authorities in the south-east and Greater London regions of the United Kingdom (Graham and Dominy, 1991). For the "peripheral" authorities, the prime benefits of advanced telecommunications are reported to derive from the technology's potential for collapsing the time and distance barriers between the local economy and the continental "heartland" firms and markets of the European Community (EC). The general implication of these geographical developments in the information economy is that ICT initiatives are increasingly directed towards improving the competitiveness of urban centres as business locations.

Types of policy initiatives

The results of the OECD surveys are a rich and unique source of information about ICT policy initiatives in the area of economic development, which local authorities are already pursuing or are planning to implement. The following discussion is intended only to illuminate the range of these initiatives through presentation of some useful examples. In the interest of clarity, and notwithstanding some overlap, the initiatives are organised by their derivative relationship to structural changes in the information economy, as outlined above: internationalisation, transaction costs, innovative capacity and education/manpower.

Internationalisation

A new and important element of local economic strategies is improving accessibility to foreign markets and industries through the shared use of international database networks. A good illustration of this is the network of *"EuroInfoCentres"* set up by the EC, whose information services are continuously updated with facts and support packages through online data links with Brussels.

The centre located in Strathclyde, Scotland, for example, provides small and medium-size firms in the region with access to:
- major EC databases on European law, R&D programmes, financial schemes and grants, and official documentation;
- an electronic mail service linking business advisers and their clients in order to generate commercial opportunities (*i.e.* markets);
- a database of local companies wishing to find European business partners; and
- the "European Contracts Information Service" which notifies local firms of tender competitions for public sector works.

Another example of local authority use of global information services is provided by the case of Turkey. Two major network services are available.

First, the TURPAK (Packet Switched Data Network) system provides users with access to worldwide online databases; it is currently established in the four largest cities (Istanbul, Ankara, Izmir and Adana) but planned to cover all provinces.

The second major network is operated by the Turkish Scientific and Technical Research Institute and provides a connection to American and European databases. Information can be obtained on a wide range of scientific and technical, educational, economic and business topics – in other words, the entire universe of information which the online industries now market on a global basis. It is planned that local authorities will become significant subscribers to these network services in the near future.

The growing international dimension of information services is most clearly revealed by intensifying local authority interest in teleport infrastructure and inter-city high-speed telecommunications links. In essence, this new infrastructure is intended to reduce the telecommunications-related locational costs of producing and distributing information services in multinational or internationally-oriented firms, as a basis for attracting mobile investment and improving the competitiveness of local businesses. For example, in Amsterdam, Netherlands, and Edinburgh, Scotland, teleport plans are aimed at attracting international head offices (corporate information services) and enhancing the global competitiveness of local financial industries (information-intensive services). In Bremen, Germany, teleport plans are part of a port information system, which is a competitive logistical tool used to improve the city's position as an international "entrepot" centre. In the Andalusian region of Spain, teleport and "electronic highway" plans are similarly aimed at increasing the global competitiveness of traditional agricultural and manufacturing industries, whilst attracting multinational firms into the area. Finally, in the Host Computer Network planned for Manchester, United Kingdom, international links to support access to global online databases and corporate information services are implemented with data communications and the worldwide GEOMAIL network of electronic mail systems (Figure 3).

Transaction costs

Recent research shows that transaction cost-minimising behaviour has become more important to the locational strategies of firms (Scott and Storper, 1987). The "information-rich" environment of urban centres, in particular metropolitan cities, has become correspondingly more influential as a locational-pull factor for mobile investment and as a basis for competitive advantage amongst local firms. A major area of local authority ICT initiatives consists, therefore, of information policies which are directed towards "subsidising" the local transaction costs of firms. These costs relate to transactions between firms, between firms and consumers, and between firms and the local authority itself.

In relation to *inter-firm information linkages,* ICT initiatives mainly involve the creation of electronic business directories or databases through which firms can search for potential suppliers and consumers. In the United Kingdom, these types of directories have been established in Coventry to promote local sub-contracting in the motor vehicle industry. A further example from Espoo, Finland, is the "service register for enterprises", a local database that lists firms and their products to encourage sub-contracting at the regional level (Helsinki region) and nationally. On a greater scale, local authorities in "entrepot" cities throughout Europe, the United States and Japan are involved in the creation of port information systems, which are intended to enhance the competitive position of these cities in international freight transport. These computer network-based systems reduce the great amount of paperwork (a significant element of transaction costs) involved in merchandise trade through electronic data interchange (EDI) – between forwarders, agents, shipping lines, banks, the port and other relevant sectors – and online information services. A recent international survey of port information systems by the Government of Canada (1990) is a useful reference source on these important policy initiatives. The INTIS Network (Figure 4), which the City of Rotterdam has helped to establish, is presented for illustration (further details are given in Annex 1).

Policy initiatives aimed at subsidising the costs of *firm-consumer* transactions are particularly important to local service industries, which are now making a critical contribution to post-industrial economic development. These types of ICT initiatives, based on the provision of online information services via city-wide computer networks and access points located in public places or tourist/travel offices, are designed to improve the consumer's knowledge (by reducing the search costs) of local facilities and services – for example, the availability of

Figure 3. **Worldwide Geomail network of electronic mail systems**

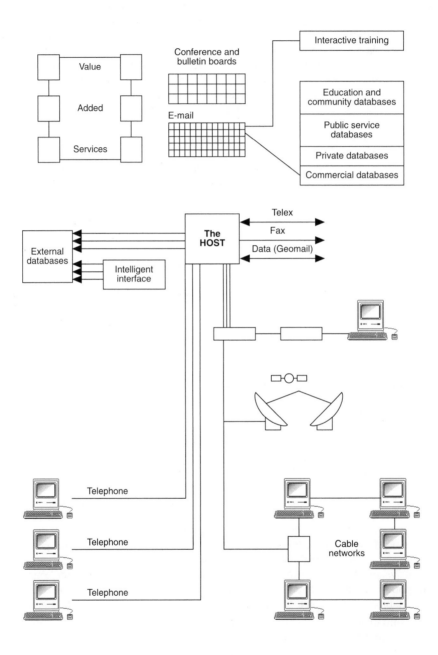

134

Figure 4. **INTIS network, port of Rotterdam**

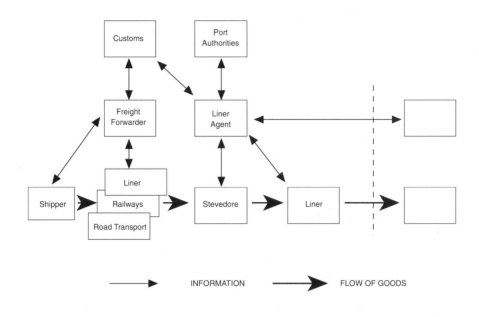

theatre tickets, bus schedules, shopping, hotel accommodation and so on. Examples of these ICT-based initiatives are to be found in the cities of Glasgow, United Kingdom; Paris, France; and Amsterdam, Netherlands. In Cincinnati, United States, the ''City Identity Kiosk'' system offers tourists a comprehensive profile of the city, including business information about restaurants, shops and services as well as a rich inventory of local history, architecture, and current public events. An important platform and catalyst for the development of these firm-consumer information services (based on advanced telecommunications infrastructure) is provided by major events – for example, the Olympic Games in Barcelona, Spain; the World Student Games in Sheffield, United Kingdom; and the Winter Olympic Games in Lillehammer, Norway.

Of further importance are ICT initiatives which reduce the costs of *firm-local government transactions*. The most widespread of these initiatives consists of offering firms access to computerised databases on planning permissions and land development rights. In addition to reducing search costs in the urban land market, these initiatives have the general effect of eliminating much of the ''red tape'' which may discourage firms from developing, locating or operating on sites in the area. Examples of these initiatives are to be found in Sheffield, United Kingdom, and Castres, France, the latter being based on a software system called the ''Communal Land Book'' which supports microcomputer access.

The vast amount of information on the local area and economy held by the municipal authorities enables them to reduce the transaction costs of firms through a variety of other ICT initiatives. Recent surveys in the United Kingdom show authorities offering computerised information services to improve the functioning of local markets for housing and labour, which influence business and staff relocation decisions, as well as social and demographic databases to subsidise the market research and advertising costs of firms. The OECD surveys indicate that these types of initiatives are extremely common in all countries. In future, they will be expanded and refined by local authority use of "Co-ordinated Information Systems" (CIS) for multi-information service delivery, as the recent OECD/Government of Denmark (1989) seminar clearly concluded.

Innovative capacity

Local authority policies to encourage product and process innovations in the economy are fairly widespread, and are generally linked to regional, national and international initiatives. A first area of policy development was discussed above – that is, authorities may act as local mediators or "brokers" and direct small and medium-size firms to scientific and technical database services.

A second area of ICT initiatives consists of technology transfer links with higher educational institutes (HEIs). These initiatives are extremely common and form key elements of "technopolis" plans in Japan, teleport plans in France, "multifunction-polis" plans in Australia, and the urban-regional programmes of the European Commission. Two examples of policy initiatives from the United States are the Ohio Innovation Exchange Network (TIE-IN) and the Illinois Resource Network: the former is a statewide interactive database that provides firms with information on university research projects, experts, venture capital opportunities and patent information; the latter is a statewide electronic directory that offers firms contacts with university experts as consultants.

The last area of ICT-based economic development initiatives is more familiar, being linked to spatial planning for high-technology industrial development – science and technology parks, at site level, and "technopolises" and "advanced information cities", at citywide level. Here, teleports and a mix of interconnected local area and metropolitan area networks are a new ingredient of technological support, with local authorities making use of their business advisory and counselling services, as well as planning, local taxation and infrastructure policy instruments to implement schemes (Williams, 1988). The technopolis programmes in Japan and France and the Australian multifunction-polis plan (based on the Japanese model) exemplify these advanced ICT-based approaches to "high-tech" development and innovation-led renewal in traditional resource and manufacturing industries. In Cologne, Germany, the city's "media park" brings together the audio-visual sector and the information industries in one planning area, which is supported by advanced fibre optic, local area computer networks. The general thrust of ICT initiatives is towards improving the innovative performance of the local economy through the systematic use of information services for spatial and functional integration between firms, government R&D/technology centres and HEIs.

Education and training

In addition to generally encouraging the use of ICT in schools and skills training centres, as a basis for improving standards of "computer literacy", local authorities are pursuing highly innovative policies in this area. In Ottawa, Canada, for example, "The Learning Centre" is an adult literacy and numeracy training drop-in establishment, which uses microcomputers and a variety of software for learner-based instruction. It is also a training centre for teachers involved in adult education and, in addition, evaluates commercial software and provides a testbed for the use of computers in literacy training. The Host Computer Network, planned for Manchester, United Kingdom, will also offer online training for commercial and public use, based on courses in, for example, microcomputer software packages which students would follow at their own pace and time. These "distance" learning or training initiatives are being organised in co-operation with the city's further and higher education institutions.

Indeed, in some OECD countries, distance learning schemes based on advanced telecommunications and cable networks are beginning to gain in popularity. The pressures for implementation may be teacher shortages, school operating costs or simple geographical remoteness. In the state of Maryland, United States, for example, teachers deliver classes to dispersed groups of students located in different high schools via cable networks and classrooms which are equipped like television studios. A similar scheme is planned to operate in the Highlands region of Scotland, where remote communities will be "wired" into teaching facilities located at the larger towns in the area. The "Infocottage" or "Telecottage" schemes being pursued in Scandinavian and other countries are more fully discussed by Lars Qvortrup elsewhere in this volume.

Perhaps the most ambitious plan for using ICT to improve the knowledge, educational and training resources of the city is being pursued in Kawasaki, Japan. Based on the "wired city" model of urban development (Blumer, Dutton and Kraemer, 1987), the City of Kawasaki plans to use advanced computer network technology to transform the area into a "university campus", with widespread online access to knowledge resources and information services being made accessible to all sectors of the local population through a number of specialised (by interest topic) "intelligent plazas". The Kawasaki "Campus City" concept is still some way from realisation; however, several basic elements of the plan are already evident in Japan and other countries. In particular, universities across the world area are initiating distance learning programmes not only within countries, but also through global teleconferencing facilities between countries.

The general picture is a future of more open-access and a-spatial educational and training institutions, whose development and resources are being increasingly recognised by local authorities.

Finally, an increasingly popular solution for firms faced by skill shortages is "telework" (International Labour Organisation, 1990). In addition to its relevance to smaller urban centres in rural regions (see Qvortrup), telework schemes implemented at the intra-metropolitan level are being used to tap the considerable reservoir of skills possessed by home-bound or less mobile groups – especially mothers with young children and disabled people. In the United Kingdom and the United States, local authorities themselves have introduced telework schemes for their employees, and, in Switzerland, these types of ICT-based work arrangements are being introduced to "take work" to remote communities in the Alpine regions. In the 1990s, telework is likely to become more widespread in OECD countries, and the current limited involvement of local authorities in these embryonic areas of the labour market will entail resolving issues to do with the flexible zoning of residential-work land uses, neighbourhood traffic generation and so on (Ruttemberg, 1988).

The institutional framework

As with other functional areas, the nature and form of local authority initiatives in the field of economic development – including ICT-based policies – vary across OECD countries owing to their different national social and economic systems. The general picture is, however, one of local authorities in all countries tending to pursue policy initiatives in collaboration with other institutions, including private businesses, regional and national governments, HEIs and other interested agencies. That is to say, authorities co-operate in networks of institutional actors in order to promote economic development in their areas, through resource-sharing and the exploitation of "local synergies" (Aydalot and Keeble, 1988).

In all OECD countries, co-operation between central government and local government occurs in several ICT application areas, with local authorities acting both as information providers and users. In Finland, for example, local authorities co-operate in a nationwide property information system (for use by banks, real estate agencies, and insurance companies), which is jointly operated by the Finnish State Computer Centre, the National Board of Survey and the Ministry of Environment. In Turkey, authorities are projected to become users of two major data communications networks established by the PTT General Directorate (Ministry of Transportation), which provide access to international databases containing scientific and technical, economic and business and other information relevant to local economic development policies. The Manchester Host Computer Network, in the United Kingdom, is being established with grants from the Government's Urban Programme. The teleport development at Metz, France, involves a public partnership between France Télécom, the Lorraine region, and the local government of the town.

In federal states, local authorities benefit from co-operation with regional government, whose ICT policy initiatives to promote economic development are also widespread. In the United States, for example, local authorities benefit from ICT-based information services which are directed towards improving technology transfer between HEIs and industry: the Michigan Technology Transfer Programme, Ohio's Innovation Exchange Network and the Illinois Resource Network are good illustrations (Schmandt and Wilson, 1990). Regional governments are also contributing to the creation of new organisational forms in the area of information services. For example, the Government of New Brunswick, Canada, has established the Geographic Information Corporation, a new agency with the responsibility and executive authority for all geographically-referenced information activities. The two major objectives of the corporation are to improve the management of natural resources and to promote economic growth through the development of marketable information products and services.

In large-scale ICT-related projects, local authority co-operation with other tiers of government is more complex and involves networking with a variety of business and educational institutions. In Germany, for example, the Karlsruhe Technology Factory project is intended to provide new firms with opportunities to develop manufacturing products and processes based on federal and local government subsidies. The project involves collaboration between Baden-Württemberg, private industry, banks, the Federal Ministry for Research and Technology and the City of Karlsruhe. In addition to universities, the project is supported by the R&D and other information services of locally-based institutions, such as the Nuclear Research Centre and the Association of German Engineers. Another example of this networking formula for local economic development – based on the systematic application of ICT infrastructure and technology transfer – is the Cartuja "93 Project under way in the Andalusian territory of Spain. Here again, improving the innovative potential and international competitiveness of the regional economy is planned around advanced technology centres located in several urban centres (*e.g.* Malaga and Cordoba), with information services and ICT (including high-speed data networks and teleports) providing the necessary linkages for spatial and functional integration.

In all OECD countries, the role of private sector and local authority collaboration is growing more significant. Private firms offer consultancy, training support, software, maintenance, turnkey projects and facilities management. In New Zealand, consulting firms have developed extensive computer databases on land use, transport, planning decisions and so on, which reinforce and complement the ICT initiatives of the local authorities. On a grand scale, the technopolis plans in Japan depend on the industrial-technological leadership of Mitsubishi Electric Corporation and the Information Network System (nationwide high-speed broadband telecommunications infrastructure for voice, data and image communications) operated by Nippon Telegraph and Telephone, supported by the Ministry of International Trade and Industry and the National Land Agency.

In the United Kingdom, teleport plans in Edinburgh and cable communications planning in all large urban centres are being implemented with a considerable contribution from the information industries (*e.g.* British Telecom, American cable companies, computer systems manufacturers and consultants) in terms of finance, expertise and management (Graham and Dominy, 1991).

In general, therefore, ICT policy initiatives in the area of economic development – and large-scale projects in particular – are pursued by local authorities in collaboration with other levels of government, HEIs and private industry. In these emerging networks of co-operative policy development and implementation, the OECD national surveys reveal a marked shift towards local authorities using ICT to strengthen their role as information generators and distributors – to encourage innovation and entrepreneurship, to reduce costs of business and to internationalise the local economy.

Assessment

As the responses to the OECD surveys showed, it is difficult to provide a reliable and measured assessment of the impacts of ICT policy initiatives in the area of local economic development for some basic reasons. First, these are early days in the innovation process and systematic effects are not discernible. Second, the "public good" characteristics of information make its specific impacts on productivity and efficiency at the level of individual firms and local authorities highly elusive. And third, the long-range benefits of ICT innovations are likely to outweigh the heavy fixed costs of equipment and organisational adjustment, which characterise the short-term and medium-term periods of the innovation process and learning curve (David, 1990).

Notwithstanding these qualifying remarks, it is possible to identify some qualitative and perceived benefits of ICT initiatives as reported in the OECD surveys.

First, the most widely reported benefit is the increased flexibility and quality of information services provided by local authorities to user firms. The general effect of less red tape and greater responsiveness (time/accuracy) is to lower business transaction costs with respect to search processes in the local labour, property and housing markets.

Second, ICT have resulted in greater information efficiency in local authority policies, producing more detailed and up-to-date knowledge of economic structures, processes and institutions. The general effect of this is overall improvements in economic development strategy and specific policy initiatives, which benefit local firms and residents.

Third, through ICT innovations, the "information brokerage" role of authorities has provided local firms with greater knowledge of potential suppliers and markets, both at the national and international level: the general result being higher levels of local investment, income and employment.

Fourth, the greater ICT-intensity of urban infrastructure is reported to have improved the locational attractiveness and marketability of sites/cities, particularly in run-down metropolitan districts and "peripheral" regions. The general implication is that ICT innovations can contribute to urban renewal or regeneration programmes and regional policy initiatives. Finally, there is supporting evidence for the view that ICT innovations can be used to increase accessibility to educational, training and work opportunities, through distance learning and teleworking schemes. The effects of this are to reduce the skills and manpower constraints on local economic development posed by locational inertia amongst firms and labour, increase participation rates amongst less mobile groups, and eliminate some of the bottlenecks in education provision.

The major constraints on the diffusion of ICT-based policies are reported to be lack of financial resources, regulatory constraints, shortage of expertise and technical barriers. The financial impediments are posed by high initial installation costs as well as the costs of labour-intensive data entry; however, the significance of these problems is expected to decrease as economies of scale are achieved and data-collection procedures are progressively automated. The traditional lack of statutory power in telecommunications has acted as a disincentive to local authorities to exploit pro-actively the economic development potential of this infrastructure (in stark contrast to transportation); however, awareness of the development potential of telecommunications is increasing amongst authorities, as computer network innovations and information services become more important to their own operations and that of local firms. As with the private sector, shortage of ICT expertise is a general problem in local authorities, suggesting that resource-sharing or pooling expertise through inter-authority links and private-public sector collaboration is both necessary and desirable. Finally, the development of information services is constrained by the technical incompatibility of different network systems used by the same local authority; again, this problem is likely to lessen with the shift towards network integration and the emergence of ISDN systems (Integrated Services Digital Network).

In the future, ICT-based initiatives in the area of local economic development are expected to grow and develop in several respects. First, local authorities will adopt more integrated, distributed processing computer networks in order to decentralise systems for delivering information services through single access points. More intelligent networks and, at the same time, the advent of voice-recognition technologies will make methods of consuming information services more simple or user-friendly. Second, the internationalisation of information services will increase considerably – especially in EC countries – as inter-urban competition for new investment and jobs intensifies on a global basis. An extremely plausible scenario is inter-authority economic co-operation of strategic alliances emerging in a global context, with ICT-based initiatives providing the information structure to support these transnational partnerships (which are becoming more common amongst private corporations). Third, authorities will place greater emphasis on improving the "knowledge environment" of the local economy in terms of ICT applications for innovation, employment training and education, and cable-based local programming to promote cultural industries. Fourth, ICT initiatives (inter-organisational computer networking) will be increasingly organised through networks of institutions to promote resource-sharing and local synergies, with authorities co-operating with a wide range of public and private sector interests. And fifth, there will be increasing trends towards charging for information services owing to pressures for new sources of revenue-generation and cost pressures, with local authorities either entering the online information business directly through new organisational forms or contracting out information service functions to specialised private firms whose price and supply schedules would be regulated by the authorities.

Policy options

Technical considerations

Local authorities should:

- conduct an inventory of local telecommunications services and evaluate the options for delivering information services to business users;
- ensure that the ICT systems for delivering information services are technically compatible with all business users, by standardising on the Open Systems Interconnection (OSI) model;
- plan for distributed network access points in their information services delivery systems, possibly using geographical information systems (GIS) technology, to ensure that services are equally accessible to all businesses in their areas;
- decentralise computer equipment, and training and technical support, to the neighbourhood or small-district level, possibly in the form of "electronic village halls", in order to make information services accessible to small businesses, the self-employed, community groups, and consumers;
- implement network access points with hardware and software which is highly user-friendly in order to make services accessible to people with limited technical skills.

Innovation diffusion

Local authorities should:

- generate publicity for their information services, through the media and by distribution of brochures, in order to increase awareness of their services;
- establish user groups amongst local businesses in order to identify service needs, generate interest and demand, and create a forum for exchanging experiences and options;
- co-operate with advanced business users of information services to initiate demonstration projects in order to highlight the costs and benefits of usage;
- establish "business information centres" in order to provide small and medium-size firms with additional technical support and commercial advice on the applications of information services;
- maximise the use of central and local libraries as "online centres" for distributing and advertising information services to community development agencies and individuals;
- promote awareness of the applications of information services (alongside IT skills training) in schools and colleges;
- evaluate the costs and benefits of charging businesses for information services (having regard to the relevant national legislation on charging practices) in terms of how price schedules may affect their rates of adoption and diffusion.

Institutional networks

Local authorities should:

- co-operate with firms, labour organisations, community groups, HEIs and other public sector agencies to produce a general policy research framework for developing the information service "environment" in their areas, paying regard to long-range planning, monitoring and co-ordination requirements and social issues;

- establish working arrangements with telecommunications carriers and computer systems manufacturers in order to optimise the planning, design and engineering of delivery systems for information services;
- establish working arrangements with HEIs in order to improve ICT-based technology transfer (such as computer-aided design or expert systems) in the local economy;
- collaborate in regional, national and international policy initiatives, operated by governments and other representative organisations, in order to maximise the spatial efficiency and effectiveness of their own information services;
- establish bilateral and multilateral collaboration by pooling information services (through reciprocal access and interconnection), as a basis for developing mutually beneficial forms of inter-urban economic co-operation at the regional, national and international levels.

BIBLIOGRAPHY

AYDALOT, P. and KEEBLE, D. (eds.), (1988), *High Technology Industry and Innovative Environments*, Routledge, London.

BATTY, M. (1989), *"Cities as Information Networks"*, paper presented at the Third International Workshop on Innovation, Technological Change and Spatial Impacts, Cambridge University, September.

CASTELLS, M. (1989), *The Informational City*, Basil Blackwell, London.

DAVID, P. (1990), "The Dynamo and the Computer: An Historical Perspective on the Modern Productivity Paradox", *American Economic Association Papers and Proceedings*, 80(2), pp. 355-361.

DUTTON, W., BLUMER, J. and KRAEMER, K. (eds.), (1987), *Wired Cities*, G.K. Hall, Boston.

GOVERNMENT OF CANADA (1990), *Port Information Systems Around the World*, Department of Communications, Ottawa.

GRAHAM, S. and DOMINY, G. (1991), "Planning for the Information City: The UK Case", *Progress in Planning*, 31(2).

HARVEY, D. (1989), "From Managerialism to Entrepreneurialism: The Transformation of Urban Governance in Late Capitalism", paper presented at the Vega Symposium, April, Stockholm.

HEPWORTH, M. (1989a), *Geography of the Information Economy*, The Belhaven Press, London.

HEPWORTH, M. (1989b), *"Local Authorities and the Information Economy in Great Britain"*, paper presented to the OECD/Government of Denmark Seminar on Co-ordinated Information Systems for Urban Functioning and Management, November, Copenhagen.

HEPWORTH, M (1990), *"Information Cities in 1992 Europe"*, UTH 2001 Series, Ministère de l'Equipement, du Logement, des Transports et de la Mer, Government of France, Paris.

HOWELLS, J. (1986), *Economic*, Technological and Locational Trends in European Services, Avebury-Gower, Aldershot.

INTERNATIONAL LABOUR ORGANISATION (1990), "Telework", *Conditions of Work Digest*, 9(1).

MOSS, M. (1987), *"Telecommunications: Shaping the Future"*, paper presented at the Conference on America's New Economic Geography, April, Washington, D.C.

OECD (1981), *Information Activities, Electronics and Telecommunications Technologies*, ICCP Series, Vol. 1, Paris.

OECD (1986), *Trends in the Information Economy*, ICCP Series, N° 11, Paris.

OECD/BRIE (1989), *Information Networks and Competitive Advantages*, (final report by Bar François, Borrus Michael and Coriat Benjamin), Paris.

PIORE, M. and SABEL, C. (1984), *The Second Industrial Divide*, Basic Books, New York.

RUTTEMBERG, R. (1987), *"Impact of Demographic and Technological Change on Urban Areas"*, report to the Expert Meeting on Urban Development and Impacts of Technological, Economic and Socio-Demographic Changes, Group on Urban Affairs, June, Paris.

SCHMANDT, J. and WILSON, R. (eds.), (1990), *Growth Policy in the Age of High Technology,* Unwin Hyman, Boston.

SCOTT, A. and STORPER, M. (1987), ''High Technology Industry and Regional Development: A Theoretical Critique and Reconstruction'', *International Social Science Journal,* 112, pp. 215-32.

WILLIAMS, F. (ed.), (1988), *Measuring the Information Society,* Sage Publications, Newbury Park, California.

WILLIAMSON, O. (1985), *The Economic Institutions of Capitalism,* The Free Press, New York.

EXAMPLE OF PORT INFORMATION SYSTEM: CITY OF ROTTERDAM

In 1982, following a study by two consultants, the city of Rotterdam, together with regional and national governments, set up a joint working group with private-sector employers and labour unions to direct a master plan for the computerisation of port activities. There was a common belief that computerisation was critical to maintaining and enhancing the Port of Rotterdam's competitive position. Rotterdam, Bremen and Hamburg are all transfer points for merchandise shipped to and from the European interior, including France, Germany, Switzerland and Belgium. In 1984, a decision was taken to implement the International Transport Information System (INTIS), at the Port of Rotterdam.

INTIS provides a store and forward message service of standard trade documents between forwarders, agents, stevedores, shipping lines, the port, customs and banks (see Figure 4). INTIS's design allows for communication of data, text, sound (voice) and images. Services provided include: core community system data communications, data translation, international telecommunications capabilities and electronic mail. It also performs document and protocol conversion between the dissimilar systems of its community of users.

The INTIS network uses facilities of the Dutch PTT Telecommunications Authority. To expand the possibilities available to users, links with CARGONAUT and SAGITTA have been established or are in preparation. CARGONAUT is the information system for the air cargo industry at Amsterdam Airport and beyond. SAGITTA is Dutch Customs' automated system for import declarations.

Many other services in the master plan for INTIS are:

- an information centre providing 24-hour INTIS support and advice;
- terminal back-up;
- a library of computer programmes (developed by INTIS or purchased, available with the INTIS communication network);
- training and user awareness (includes initial training, courses for users, *etc.*);
- a transportation monitoring management system;
- an electronic monitoring information system (linked to financial and community exchanges);
- a supply inventory management system;
- access to databases – company, tariffs, regulations, ship movement, Lloyds and other ship registers, registration, transportation routes, *etc.*; and
- data processing services.

INTIS has created gateways to the international computer networks IBM and GEIS. It is now possible to set up links with other port systems such as Felixstowe's FCP80 (described later in this report). FCP80 is connected through TRANSPOTEL (developed by the United Kingdom, the Netherlands and West Germany, and by a group of freighting experts in Switzerland) to European trading partners. It indicates who in the United Kingdom and Europe is moving freight and who has freight capacity available.

One of the distinctive features of INTIS is that it supports the EDIFACT standards for data and message formats. This will enable INTIS to connect easily and cheaply (since not every connection will require customised programming) with its users and with other systems worldwide.

Formally incorporated as INTIS in August 1985, the initial shareholders were the City of Rotterdam (including the port authority) at 40 per cent, the PTT at 9 per cent, private companies at 40 per cent, and others at 11 per cent. Sales of shares plus initial capital development grants from the Dutch Ministry of Economic Affairs Services provided start-up financing for the project.

TELECOMMUNICATIONS
AND URBAN ECONOMIC DEVELOPMENT

Mitchell MOSS
Director of the Urban Research Center
New York University, UNITED STATES

*Mitchell L. Moss is professor of Public Administration and Planning and direc-
tor of the Urban Research Center at New York University. He has served as a
deputy of the chairman of the New York State Council on Fiscal and Economic
Priorities and as chairman of the Interactive Telecommunications Program at
New York University's Tisch School of the Arts. He received his B.A. from
Northwestern University, M.A. from the University of Washington, and Ph.D.
from the University of Southern California.*

*Professor Moss has directed projects on telecommunications and regional
development for numerous public and private organisations and currently
serves on Mayor David Dinkins' Task Force on Telecommunications (New York
City). His articles include "Telecommunications, World Cities, and Urban Policy" (Urban Studies), "Public Policy and Community-Oriented Uses of Cable
Television" (Urban Affairs Quarterly), and "The Diffusion of New Telecommunications Technologies" (Telecommunications Policy). He is the editor of Telecommunications and Productivity.*

INTRODUCTION

Telecommunications technologies present challenges and opportunities to cities and local
governments concerned with fostering economic development.

Urban growth historically has been closely linked to the development of critical public
infrastructure such as public markets, maritime facilities, railroads, industrial parks, airports,
and roads and highways. With the rise of information-intensive activities in both the service
and manufacturing sectors, the flow of information in, through and out of cities has emerged as
a vital element in the urban economic and social system. Building an infrastructure for the
movement of people and goods is not enough to assure urban economic development; it is also
necessary to have a modern telecommunications infrastructure that can allow information flow
in, through and out of cities and regions.

Unlike many urban elements of the urban infrastructure that were initiated by cities – such
as mass transit, port development, and water supply – policies for the telecommunications

infrastructure have traditionally originated at the national level with virtually no involvement of cities and local government. "The predominant role of either regulated monopolies or central government agencies in the design and development of communication facilities has meant that there have been few opportunities for local governments to intervene, whether through land use regulation or capital assistance"[1].

With the privatisation of once publicly-owned telecommunications carriers, the deregulation of public monopolies, and the rise of competition in the telecommunications industry, local and regional governments have become actively involved in using information and communications technologies to foster economic development. The emergence of deregulation and advances in the telecommunications industry pose a special challenge for cities since telecommunications technologies can have the dual effect of fostering both the centralisation and decentralisation of economic activities.

Communications technologies have made it possible for firms to co-ordinate and control highly dispersed activities – manufacturing, assembly, and data-processing – from a single point, typically a headquarters located within a large central city. Telecommunications also facilitates the decentralisation of activities by allowing firms to move functions far from headquarters, but at locations which offer access to low-cost labor, raw materials, or proximity to markets. Further, telecommunications systems are widely used in the sectors of the urban economy that are growing most rapidly: business services such as finance, advertising, media, and management consulting. As the world economy becomes more highly integrated, telecommunications will allow firms to develop new services and reach new markets, and this should contribute to the long-term, economic development of major cities.

In considering the relationship of telecommunications to urban economic development, it is essential to recognise the growing role of information and communications technology in office activities, the design of office buildings, and in the global operations of private firms. It is also important to develop an awareness of the criteria that determine private sector investment in telecommunications infrastructure so that local governments can attract such investment and link it to urban development priorities. Further, cities should use telecommunications to improve public sector productivity, and thereby stimulate a local market for telecommunications vendors and the organisational capacity to apply technology to economic development programmes. This paper identifies four areas in which cities are using information and communications technologies as part of their economic development strategies:

- the development of teleports and high tech office centres;
- local policies to encourage private investment in telecommunications infrastructure, *i.e.* intelligent buildings, and value added networks (vans);
- public policies to use technology in public services that affect economic development;
- local policies to attract information-intensive firms and activities.

Types of applications

Teleports and high tech office centres

The widespread use of computers in office functions and the need to have access to advanced telecommunications systems has generated a demand for office buildings designed to serve the specialised needs of information-intensive firms. Technology has dramatically altered the infrastructure of office buildings, leading to the need for large floor areas capable of accommodating the modern "trading floor" as well as advanced computer systems. The spectacular success of La Défense in Paris demonstrates the way in which new business activity requires office buildings with sophisticated communications services, and the potential for locating these activities next to each other, outside the traditional central business district of a city.

The Infomart at La Défense is especially notable for it brings together numerous information-related firms in a single place to market and display their products near the firms that are most likely to use them.

As the report, *Teleports in a European Context* points out, there is substantial activity to build teleports throughout western Europe, with teleports planned or underway at Ile-de-France, Roubaix, Metz, Avignon and Nice. The Ile-de-France teleport will serve La Défense plus the Paris stock exchange and related financial services. Teleparcs will also be built that are linked to the Teleport to provide advanced services in other areas surrounding Paris. The Ile-de-France teleport represents an innovation in institutional co-operation among developers, the Chambers of Commerce and Industries of Paris and Ile-de-France as well as cable network operators.

The success of the teleport in Staten Island, New York has encouraged the development of similar teleports in cities as far from one another as London, Rotterdam, Amsterdam, Tokyo and Osaka. In the United States, most teleports are privately financed and consist solely of communications equipment linked to satellite and local terrestrial networks; users are not physically located at the site but gain access to the teleport through a regional distribution system.

By contrast, "real estate-based teleports", similar to the one on Staten Island, New York have been developed in San Francisco and Dallas and in cities outside North America. They constitute a way to use high technology to stimulate real estate development.

In Tokyo, a teleport is to be built as part of the long-term offshore development programme of the Tokyo Metropolitan Government on a 98 hectare parcel of reclaimed land in Tokyo Bay located six kilometers from central Tokyo. The Osaka Teleport – already in operation – is a main element in the project known as Technoport Osaka supported by the "Model Cities" programme of Japan's Ministry of Post and Telecommunications which is designed to reclaim land from the harbor and modernise the city's trade, transportation and telecommunications infrastructure. The Osaka Teleport's satellites serve Pacific and Indian Ocean satellites and are to be linked to the greater Osaka with an integrated digital fibre optic network being built by the Osaka MediaPort Corporation.

The London Docklands Development Corporation which is overseeing the revitalisation of the once run-down and decaying docklands of London has also included satellite earth stations in its overall development project. The teleports – operated by Mercury Communications and British Telecom – will be in close proximity to Olympia & York's Canary Wharf development, which consists of 24 buildings on 71 acres including a 50 story skyscraper that will be the tallest building in the United Kingdom when finished. According to the London Docklands Development Corporation, the presence of the satellite earth facilities on the Docklands provides a high-tech image for this area and reinforces efforts to attract financial service firms and communications companies to this site, which competes for offices with other locations in London and the southeast of the United Kingdom.

In Rotterdam, the world's largest port, a teleport has been built in conjunction with the port, so as to provide advanced communications services for all shipping and trade functions. The growing use of communications satellites to guide and co-ordinate shipping – through

MARISAT – highlights the way in which transportation activities and telecommunications systems are closely linked and are an integral element of modern port functions.

Further, these examples of teleport development demonstrate the potential role of local government in providing the land for satellite earth stations and in using teleports as a magnet to stimulate office development by the private sector.

Policies to encourage private investment

The telecommunications infrastructure of the future will be predominantly built by private sector firms; however local governments can and should play an active role in encouraging private investment in this infrastructure by making rights-of-way available for fibre optic systems, by providing incentives so that the infrastructure serves all parts of a community, and by identifying opportunities for public-private cooperation.

Governments control the rights-of-way on highways, mass transit systems and railroad lines. These rights-of-way are of great value to firms seeking to build fibre optic networks since they link major cities which continue to be the major sources of information traffic. Moreover, once permission to utilise a public right-of-way is granted, construction can be completed in a timely fashion since the corridor is already clear and accessible. MCI Communications has installed its fibre optic network along the AMTRAK rights-of-way between New York and Washington; and Metropolitan Fiber Systems, a provider of local fibre optic networks in major US cities, has installed fibre optic in a conduit leased from the Southeastern Pennsylvania Transit Authority and adjacent to Philadelphia subway tunnels. One benefit of underground rights-of-way is that the fibre is well protected and less likely to be damaged by street construction.

In order to install a fibre optic network in the City of London, Mercury Communications relied on the old and unused rights-of-way of the hydraulic system that was far more accessible than the narrow and highly-congested street grid of the City. In Italy, a fibre network is being built along the country's 5 000 km highway system; it will link more than 300 toll stations. Although the network is principally designed to serve administrative functions, it demonstrates the way in which transportation rights-of-way can be used for information and communications systems.

The importance of providing adequate telecommunications capability in urban areas seeking to attract new economic activity is reflected in the decision by the London Docklands Development Corporation to install 30 km of fibre optic cable in the Docklands that replaced obsolete cable. By providing ducts for the cable during the construction of new roads in the Docklands, the London Docklands Development Corporation highlighted the benefit of coordinating the construction of all new infrastructure – roads as well as communications – prior to development. An effort to assure that the Scottish Highlands have access to the latest telecommunications infrastructure is underway to assure that economic development initiatives can utilise telecommunications in this remote area.

As information and communications technologies have pervaded office functions, they impose new requirements for buildings that can accommodate the energy, space and telecommunications needs of the modern office. In New York City, the older, narrow buildings in the Wall Street financial district have a vacancy rate twice that of newer office buildings because of their limited capacity to provide sufficient energy for cooling and duct space for telecommunications equipment. Real estate developers are increasingly providing the capacity for advanced telecommunications equipment as well as higher energy loads for cooling and operating equipment.

An "intelligent" or "smart" building can consist of an automated management system for heating, ventilation, elevators, security, and lighting; special facilities for information and communications equipment such as raised floors, vertical duct space, and dual entry points for telecommunications lines; and "shared tenant services" where tenants can obtain telecommunications services from a vendor chosen by the developer. Although the marketing of "smart buildings" has had limited success[2], there continues to be a strong need for office buildings that can accommodate advanced communications systems. "The more information is handled, stored and retrieved electronically the more vital it is that buildings have the capacity to accommodate IT (information technology). Buildings have become, in a sense, an extension of the computer"[3]. In Canada, an organisation devoted to automated buildings has recently been created, the Canadian Automated Building Association, that includes representatives of business and government, and is intended to assure compatibility in regulations, standards, and building codes.

Government efforts to develop new telecommunications networks have traditionally been initiated to spur economic development through the sharing of technical information among universities, government, and industry. Michael Batty has analysed the evolution of computer networks for academic and research purposes and has wisely noted the importance of the linkages between local area networks and national networks. "What is manifestly clear in the study of computer-communications networks is that a high degree of central coordination and planning is required to enable subnetworks to communicate with one another and to enable a common infrastructure to be developed for diverse applications"[4].

The deployment of computer networks is especially significant in the financial services industry where stock exchanges – once anchored to specific places in major cities – have become electronic networks linked to buyers and sellers around the world. Although local governments do not have authority over financial service activities, cities cannot afford to ignore the economic development implications of such computerised financial service functions. Several cities have sought to establish cooperative ties among their stock exchanges in order to enhance their competitive position with the larger exchanges in New York and London. For example,the Stockholm Stock Exchange is attempting to create a Nordic exchange by developing point-to-point and satellite connections with the Oslo, Helsinki and Copenhagen markets. At present, trading volume in many Swedish companies is higher in London than in Stockholm, and this joint venture represents an effort to attract business from other large European centres.

Technology, public services, and economic development

Local government is a major user of information and communications technology. Many public services are information-intensive, such as tax and finance administration, land use zoning and regulation, libraries, education, and criminal justice. The largest user of telecommunications in New York City is the municipal government, yet there is remarkably little coordination among the various city agencies and regional authorities. Public sector organisations constitute a significant market for information and telecommunications equipment, yet they rarely use their market power to achieve economic development objectives. For example, cities could link procurement policies to locational decisions, to manpower training programmes, and to providing communications services that enhance the quality of urban life.

Public uses of information and communication technologies can have considerable economic development benefits. For example, efforts to apply technology to public services can

151

help improve the skills of municipal employees, thus enhancing their own productivity. In addition, information and telecommunications firms will often locate near their large customers in order to gain access to market information and to respond to customer preferences.

The port of Bremen was one of the leaders in using electronic data processing on a shared basis with more than 100 customers. The port has created a separate firm, Datenbank Bremische Haefen which operates the Logistics Information System of the Bremen ports, providing handling and accounting services as well as information on ship schedules and access to international trade and financial databases. This firm has recently expanded its services to include advanced telecommunications services more cost-effectively than the individual firms can do themselves.

One of the most successful applications of computer-based communications to economic development has been through the state-run computer lotteries in the United States. These lotteries, which are operated in 40 of the 50 States, provide revenue for public services without adding to the tax burden of individuals and firms. In France, a betting system run by five French horse racing associations uses the France Telecom system to provide 6 500 terminals in cafes for the registration and payment of bets.

Telecommunications can also be used to provide tourist information as is currently being done on the French Minitel system and in hotels around the world. It is remarkable how many hotel and car rental firms use ''800'' numbers to serve customers, but that public agencies have only recently begun to market their tourist attractions with such free long distance calling services. As cities begin to recognise the economic contribution that tourism makes to urban economic activity, it will be necessary to use computer networks and videotex services in their promotional efforts.

Policies to attract information-intensive activities

Many cities have come to recognise the importance of information and communications firms in shaping their economic development policies. In Cologne, the city has created a ''MediaPark'' in the inner city as part of a programme to strengthen its position as a centre for broadcasting and publishing. In Japan, there have been numerous efforts to harness information and communications technology to urban development. According to Newstead, ''there is an almost obsessive attachment to the concept of information cities, which feature in the national plans of central government agencies and in the strategic plans of municipalities'[5]. Over the past decades, there have been a series of programmes that emphasize telecommunications and economic development, encompassing two-way cable television, fibre optic networks, videotex, and teleports. Batty summarises this national and local policy well: 'Over the last decade, Japan has pursued the development of information technology with a single-mindedness and zeal which cannot be surpassed by any other nation state. This is manifest at all levels in the belief that information technology will enable economic and social solutions to many of the deep seated paradoxes facing modern Japan'[6].

One example of this emphasis on information and communications technology is the decision by Kawasaki to become an information city' by creating 18 ''intelligent plazas'' that will be nodes of smart buildings linked by a 30 km optical fibre system. Although implementation of this plan is not certain, the fibre system is being installed and other efforts to strengthen Kawasaki's downtown are underway as well. Approximately 240 000 workers commute to Tokyo from Kawasaki each day, and one goal of the Kawasaki plan is to stimulate neighborhood offices at each of the 18 "intelligent plazas" by relying on advanced communications systems.

In August 1990, The Mayor of the City of New York established a new task force to focus on telecommunications policy issues identified in a report issued by the New York City Partnership, "The $1 Trillion Gamble: Telecommunications and New York's Economic Future". Among the findings of this study were that more than 35 per cent of all the international calls from the United States originate in New York City and that New York State residents and businesses make more calls each year than are made by all the residents and businesses in France or in the United Kingdom. The report highlighted the fact that New York had 470 office buildings served by optical fibre, compared to 110 buildings in Tokyo and 400 buildings in London.

Institutional analysis

The discussion of the applications in this paper highlights the diversity of organisational arrangements used to foster economic development through telecommunications. Given the dynamic role of the private sector in developing the emerging telecommunications infrastructure, it is essential for cities to deploy a variety of strategies and tools. At the minimum, cities should monitor developments in the telecommunications industry, with a special attention given to assessing the impact of new technologies on their locational advantages and assets. Where new telecommunications systems have been built by the public sector, it has typically occurred through a joint venture, in which public agencies stimulate private sector initiatives, as occurred with the Teleport in Staten Island and the London Docklands Development Corporation.

Cities are best able to directly use telecommunications when it can strengthen or improve their existing economic development programmes, such as assuring that fibre optic systems are provided in development areas.

Furthermore, land use controls and tax policies that encourage technology-intensive office buildings are essential to assure that information-intensive firms locate in urban areas where space is costly and regulations intense.

Assessment

It is too soon to evaluate the impact of the information and telecommunications systems described in this paper. Technological innovations require time to diffuse; for example, it took the telephone more than 50 years to reach 30 per cent penetration of American households. Furthermore, advances in the technology of information and communications have been superseded by the rapid pace of liberalisation, privatisation, and competition that now characterises the telecommunications industry. It is possible to identify emerging trends in telecommunications that cities and local governments should address.

First, the growth of new communications technologies has not eliminated the face-to-face activities that occur in central cities. There is no substitute for direct personal contact for the exchange of confidential, specialised, and timely information. In fact, as more information becomes computerised, the value of face-to-face transmission of information is enhanced since there are fewer opportunities for such specialised and personalised information to be exchanged.

Second, the internationalisation of economic activities will require cities to make certain that advanced, competitive telecommunications services are available to the firms located in a given community. Regulations that restrict communications innovations and that create pricing schemes that reduce the use of telecommunications will inevitably endanger a community's economic development. Thus cities need to monitor telecommunications policies at the national and international level in order to maintain their competitive position as centres for face-to-face and electronic communications.

Finally, cities should be prudent risk-takers in the deployment of information and telecommunications technologies in their own organisations. ''These technologies offer opportunities for accelerating economic development, for facilitating economic transactions, for improving the quality of life, for establishing new patterns of social interaction.

''At the same time, they allow urban governments to develop new, more flexible services and to make more efficient the functioning of urban administration''[7].

POLICY OPTIONS AND RECOMMENDATIONS

Cities are natural laboratories for technological innovation, and policies to encourage new uses of technology at the local level are essential as are systematic efforts to share knowledge, lessons, and experiences among cities and their leaders. Furthermore, the development of information and telecommunications technologies by the private sector is proceeding rapidly and cities must develop ways to draw upon private sector innovations so as to improve the performance of government operations especially in areas related to economic development. Most important, the challenge of telecommunications lies in the need to recognise the relationship between technology and the fundamental elements of the urban infrastructure: transportation systems for the movement of people, urban amenities that foster civic and cultural life, and real development that accommodates advanced business services. It is far more important to understand the way in which information and communications technologies impinge on everyday aspects of urban life than to focus on a specific technology.

NOTES AND REFERENCES

1. MOSS M. L. (1987), "Telecommunications, World Cities, and Urban Policy," *Urban Studies,* 24, p. 534.
2. CARLINI J. (1990), "What Ever Happened to Intelligent Buildings?" *Telecommunications,* February.
3. DUFFY F. and HENNEY A. (1989), *The Changing City,* Bulstrode Press Ltd., p. 33.
4. BATTY M. (1989), "Cities as Information Networks: The Evolution and Planning of New Computers and Communications Infrastructures", Paper presented at the Third International Workshop on Innovation, Technological Change, and Spatial Impacts, Cambridge, September.
5. NEWSTEAD A. (1989), "Future Information Cities," *Futures,* June, p. 263.
6. BATTY, p. 20.
7. FOX-PRZEWORSKI J. (1989), "Information and Communication Technologies: Are There Urban Policy Concerns?" Paper presented at the international symposium on Communication and Spatial Communications, International Geographical Union, Geneva, November, p. 19.

Annex 2

CASE STUDIES OF NEW YORK, METZ AND MONTREAL

Case study: the Teleport of New York City

The original idea for a teleport was developed by the Port Authority of New York and New Jersey, a bi-state agency created in 1921, which operates maritime facilities, airports, bridges, and tunnels in the New York Metropolitan Region. In the late 1970's, the Port Authority began the planning for a "teleport" – modelled after an airport – in which a large parcel of land would be set aside for satellite dishes operated by different communications companies, with a fibre optic system providing direct access to the teleport, thereby overcoming the microwave congestion of New York City. The actual development of the Teleport involved two units of government: the City of New York, which owned a 350 acre site on Staten Island that was relatively free of radio interference, and the Port Authority, which leased land for forty years from the City of New York. A private firm, Teleport Communications, Inc., – now 97 per cent owned by Merrill Lynch – was created to develop and operate the telecommunications facilities and fibre optic network. From the outset, the Teleport consisted of three key elements: a regional fibre optic system, a multi-satellite earth station facility, and an office park for telecommunications-related businesses.

There are currently 20 earth stations operating at the Teleport, which has a capacity for 30 earth stations. Satellite transmission is managed by IDB Communications Group, Inc. and is used by broadcast networks such as ABC, CBS, and the European Broadcasting Union as well as major financial companies. The regional fibre system extends for 174 miles, serves 247 customers in 162 office buildings, and has become a competitor to New York Telephone for local high-speed business communications. Real estate development is under the jurisdiction of the Port Authority, and more than one million square feet of office space has been built at the Teleport; tenants include Merrill Lynch, Recruit USA, Nomura Communications, and Dun & Bradstreet.

Case study: Metz Teleport

The Metz Teleport is designed to stimulate regional economic development and involves France Telecom, the Lorraine Region, and the City of Metz. The Metz Teleport has been in use since 1988 and uses voice and data to serve a variety of clients. The services offered by the Metz Teleport include video conferencing, telemarketing rooms, access to international carriers on five continents, and advanced data services.

Among the facilities at Metz Teleport are a PABX with a capacity of 3 000 lines transmitting voice and data; a local fibre optic network enabling clients to be directly connected to the Teleport, and an earth station for satellite communications that reaches all French cities plus other European cities.

The City of Metz financed the infrastructure of the Metz Teleport which occupies 240 acres of land. Since its inception, it has attracted 80 companies on the site already, providing 1 300 jobs.

156

Case study: Montreal Teleport

The Montreal Teleport is located near Montreal Center and close to communication companies, such as Radio-Quebec, Videotron, and Radiomutuel. It occupies less than 13 400m². The institutional supporters of the project are: Telesat Canada and SGD (Management and Development Company).

Telesat and SGD have attracted tenants involved in programming, production, computer and telematic systems, and computer distributors. The companies located at the Montreal Teleport are attracted to the direct satellite access plus the fibre optic network that links all the buildings at the site.

INFORMATION TECHNOLOGIES AND HEALTH CARE SERVICES

Neil WRIGLEY
Professor at the University of Wales (Cardiff)
Department of City and Regional Planning
UNITED KINGDOM

Neil Wrigley holds a B.A. from the University of Wales, a Ph.D. from Cambridge and a D.Sc. from University of Bristol. Until April 1991 he was professor and head of the Department of City and Regional Planning at the University of Wales, Cardiff. Today he holds the chair in Geography at the University of Southampton. Since 1984 he has served as committee member of the UK Economic and Social Research Council. He chairs the University Advisory Council for the 1991 UK population census.

The author of 8 books and almost 100 journal articles, he is best known for his work in research methods and quantitative social science, where his books include Categorical Data Analysis (Longman, 1985) and important contributions to longitudinal data analysis, discrete choice modelling, categorical data analysis, and diagnostic techniques. His interests in health care ICT date from the mid-1980s when he developed the first operational health care planning GIS for the City of Bristol.

INTRODUCTION

In all OECD countries, health care services consume a significant proportion of GDP. In 1987, for example, total health spending in OECD countries averaged 7.3 per cent of GDP and ranged from 3.5 per cent in Turkey and 5.3 per cent in Greece to 9 per cent in Sweden and 11.2 per cent in the United States. Moreover, this proportion has been increasing steadily over time. In the period 1960-75, for example, per capita health spending grew on average 17 per cent faster in real terms than per capita GDP, whilst in the period 1975-87 it grew 13 per cent faster, increasing the average OECD-state total health spend from 6.5 per cent of GDP in 1975 to 7.3 per cent in 1987[1].

Expenditure on information and communication technology (ICT) as yet forms only a relatively small proportion of this total health care spend, but it is a proportion which has been increasing rapidly over the past decade. In the United Kingdom, for example, ICT expenditure in 1988-89 was approximately £130 million or 1 per cent of the National Health Service budget, but the UK Government has recently indicated its strong support for the expansion of ICT within the health service and, in November 1989, it announced the allocation of an extra £103 million. This cash injection will serve to bring the United Kingdom much closer to the average European level of approximately 2-3 per cent of total health care expenditure devoted to ICT. But European OECD countries are still some way behind the United States which, by the late 1980s, was estimated to be allocating more than 6 per cent of its total health care expenditure to ICT[2].

The purpose of this paper is to review the current status of ICT in the provision of health care services in OECD countries and, in particular, to consider the inter-relationship between health care ICT and the delivery of health care services to the populations of urban areas. Understanding the nature of that inter-relationship involves, in turn, an appreciation of the way in which governments in several OECD countries are seeking to transform existing public health financing and delivery systems into more market-based systems of "managed competition". It will be shown that ICT is the means by which a market-based system of health care provision can be put into operation, and that potentially this has important consequences for integration of the responsibilities of health care "purchasers" and urban government.

The structure of the paper is as follows:

First, the main objectives for which ICT systems have been used in health care provision will be outlined and placed in a context of ever-rising health care costs, restricted budgets, and the search for efficiency.

Next, the range and types of application of ICT systems will be reviewed and illustrated with case studies. In particular, attention will be concentrated on the monitoring and management of public health in urban areas via health care GIS systems, and on experiments in urban areas with smart card systems of patient-held medical records.

The third section will discuss the institutional context of expanding ICT provision and, in particular, will consider the crucial role which ICT systems are expected to play in facilitating the transformation of public health care provision systems towards more market-based systems of managed competition. The example of the reform of the NHS in the United Kingdom will be used to illustrate this theme.

The concluding section will discuss the urban government context of health care ICT and list some of the policy options open to urban government within the context of emerging, heavily ICT-dependent, market-based systems of health care provision.

Objectives in the use of ICT systems in health care provision

OECD countries operate a very wide range of health care systems. At one end of the spectrum, the United States operates a decentralised, pluralistic and competitive market-based system in which private health insurance and private-sector health-care delivery are dominant and in which the public share of total health expenditure is only 40 per cent. At the other end of the spectrum are national collectivised public provision systems, such as those operated in the United Kingdom and Sweden for much of the post-war period, in which health care has largely been free at the point of use (or available for a nominal charge) and subject to centralised political direction and planning. In between these extremes lie the majority of OECD countries which operate some form of comprehensive social-insurance-based health care system, provide about 75 per cent of health care expenditure from the public purse, but have a greater degree of diversity and decentralisation in the national system of provision than is the case in Sweden and the United Kingdom. Overlaid on this spectrum of systems there is also the important trend within many OECD countries for governments to seek transformation of existing health care financing and delivery systems into more market-based systems of managed competition. These seek a middle course, attempting to capture some of the advantages of free markets without their disadvantages, and to promote efficiency whilst safeguarding equity[3].

Clearly, these variations in the types of health care systems condition the nature of ICT take-up and usage. However, despite the considerable variation in the types of health care systems operated in OECD countries, the pressures which have resulted in the adoption of ICT have many similarities across member states. In particular ICT systems appear to have been used to address five main objectives.

Administrative efficiency and the monitoring of resource usage

The rising cost of health care provision is a common problem for all OECD countries, and is a particular concern for those countries with an ageing population (due to low birth rates in the 1970s and increased life expectancy) which can be expected to make a significantly greater demand upon both acute and non-acute medical services in the early twenty-first century. ICT has been introduced, therefore, against a background of restricted budgets and the need for cost savings and efficiency of provision. Much of ICT in the health care field reflects this ethos. In particular, hospital information systems (HISs) which dominate the total ICT expenditure within health care services are, in many cases, as much or more to do with the reduction of administration costs (particularly in the US context of automated billing systems and where no less than 22 per cent of total health expenditure is estimated to be devoted to administration)[4] as to do with direct management and improvement of patient care.

Traditional versions of these systems (*e.g.* the HAA and PAS systems in the United Kingdom – see below) have been operated essentially as paper record replacement systems, to provide information on overall hospital activity levels and resource usage, together with information on the progress of each individual patient through the hospital system. Moreover, they have essentially recorded information on what has happened in the past. More recent HISs, however, have focused on ''concurrent monitoring'', allowing clinical records/findings on each patient to be accessed within the hospital ward or at the patient's bedside and prompting particular courses of action whilst the patient is still under care. Most importantly, such systems allow evaluation of resource consumption during the in-patient stay, and the patient's progress can be monitored to ensure that bottlenecks in laboratories, operating theatres *etc.*, are not adversely affecting the treatment and unnecessarily[5] prolonging the length of stay – hospital throughput being a critical aspect of increased productivity in the delivery of health care.

Smaller versions of these HISs are increasingly being used by individual doctors (local practitioners/GPs) in urban areas or by groups of local practitioners. The pattern of usage of these systems has followed that in the hospital sector but, as might be expected, there is a much wider range in ICT take-up and usage amongst local practitioners than in the hospital system. The current position ranges from the individual doctor with no ICT support, paper records and traditional patient management procedures, through the individual doctor with a PC and a rudimentary database management system, to group practices with local networks, patient-held medical records (PHMR) systems in the form of simple magnetic-strip cards or (particularly in France) smart cards, and linkage to wider regional medical or hospital information systems. The leaders in the use of ICT in the local practitioner sector are acknowledged to be the multi-speciality prepaid group practices in the United States (*e.g.* Kaiser Permanente, Harvard Community Health Plan, Group Health Co-operative of Puget Sound). Because of the system under which they operate, of fixed prepayment by patients, these practices have the incentive to solve patients' medical problems efficiently whilst holding down costs. In such a system, making the correct diagnosis promptly and treating the patient without causing complications is rewarded (in contrast to a fee-for-service system in which more visits mean a greater fee). As a result, these practices have been leaders in the use of ICT for systematic quality measurement and control, and they have orderly processes for technology assessment and organised responses to changes in technology[6].

Displacement of non-acute cases from the acute health care system

The rising cost of acute health care and the need for efficiency of provision place a premium on removing non-acute cases from the acute health care system. ICT has a very important role to play, therefore, in keeping elderly or handicapped people out of the hospital system and providing them with the functional support necessary to remain largely self-sufficient in their own homes.

In this context, there is a great deal of interest in the concept of the ''interactive home'' which is planned to help the elderly or disabled housebound person combat the problem of isolation. Such homes would be part of a generalised communications system and would be fitted with a transmitter associated with a Minitel plus a portable trigger in case the person needed help and was unable to move for some reason. The control centre would have a host computer and video monitor, operating on a packet switching telephone network or a digital network.

The aim of such homes is to promote a better flow of information between the services provided at the home (domestic help, district nursing *etc.*) and the other people involved in supporting the housebound patient (family, neighbours, doctors *etc.*), and also to allow rapid and appropriate intervention in the case of an emergency[7]. Within such homes, the elderly or handicapped person could also help combat isolation by becoming part of a telecommunications-based network service which would:

- facilitate the exchange of mail;
- have specialised electronic information/bulletin boards; and
- allow electronic phone conversations.

Disability Information Services of Canada (DISC) provides an example of such a network service.

The measurement and management of health outcomes

In many OECD countries, systematic recording takes place of a selected number of key pieces of information on all hospital in-patient cases – usually within an HIS – and that information is reported to a central database. These so-called uniform hospital discharge data sets can then serve as a foundation for measuring what are referred to as *risk-adjusted measures of outcomes (RAMOs)*. RAMOs compare the number of actual adverse medical outcomes (*e.g.* the number of perinatal deaths or mortalities from a particular type of elective surgery) for particular hospitals, against the expected number of adverse outcomes calculated on the basis of a suitable risk-adjustment model which takes into account factors affecting the risk of adverse outcomes and the characteristics of the population at risk. Hospitals which are found to have significant differences between actual and expected numbers can then be subject to further investigation.

In the United States, risk-adjusted mortality studies of hospitals within particular states are increasing in number, and are revealing wide variations in performance. In California, for example, RAMO studies have shown significant variations across hospitals in the risk-adjusted perinatal mortality rate[8] and in the risk-adjusted death rate of heart bypass patients[9]. Moreover, the names of hospitals with risk-adjusted mortality rates significantly above or below average have been published[10].

RAMOs potentially provide hospital managers and other health care administrators with the "added-value" information required to evaluate the effectiveness of the treatment being offered by particular parts of the health care delivery system. In theory, at least, they allow administrators to identify problem areas, to make adjustments to the delivery system, and to raise standards. In other words, RAMOs can potentially be a very useful tool in the management of health outcomes. If they are published, consumers (actual or potential patients) can, theoretically at least, make informed and rational choices within the constraints of the overall health delivery system of that country. Individual choice and system management can thus become subtly interlocked[11]. The choices (and political pressures) of better informed consumers/patients should help to keep the health care managers up to scratch, whilst the improvements in management ICT systems which result from adding in the results of RAMO studies not only provide the managers with the necessary tools for improving the health care delivery system but also, in due course, can become the source of improved information for the potential patients.

The monitoring of public health

In addition to providing acute and non-acute medical care, health care managers in many OECD countries are also charged with monitoring the general state of public health in regions, in smaller health districts, or in individual urban areas. This task usually involves relating various measures of mortality and morbidity (notifiable diseases, cancer registrations, congenital malformations *etc.*), plus the take-up of immunisation and vaccination services, to population census information on the demographic structure of the population in the particular area being monitored – the aim being to obtain standardised rates and ratios. For example, in many OECD countries national or regional Cancer Registers have been established. The Standardised Registration Ratio (SRR) for any particular type of cancer (stomach, colon, breast *etc.*) in a particular health district is then the number of observed cancer registrations in that district divided by the number which would be expected if the national average age-specific registration rate applied to that district.

Traditionally, public health reports produced from this type of monitoring[12] have treated the health district as a single unit and have compared measures of public health in the district against regional or national norms. In recent years, however, there has been increasing interest in monitoring and assessing *within-district* geographical variation in public health. A particular type of cancer (*e.g.* leukaemia) or a particular notifiable disease (*e.g.* meningitis) might show considerable variation in its level of occurrence across the census tracts within a single urban area or health district. It is then important to assess whether this is "real" or apparent variation. And, if it is deemed to be real variation, it is then of interest to consider the geographical co-variation of this indicator of morbidity or mortality against socio-economic variables abstracted from the census, or against a wide range of environmental measures (*e.g.* background air pollution, heavy metal contamination of soil or water, toxic discharge from industrial plants or inefficient waste incinerators, and so on). Such investigations of the geographical co-variation of morbidity/mortality measures and socio-economic/environmental variables can, in turn, help identify relationships and/or small geographical areas which might repay further investigation in the form of carefully designed, in-depth, epidemiological studies.

The ICT tool which has, in recent years, revolutionised such monitoring of public health is the Geographical Information System (GIS). However, as we will see further on, the GIS products which have been used in public health monitoring have typically not been the large "turnkey" proprietary systems found in the utility companies (gas, electricity, water) or in natural resource management. Rather, they have been smaller "customised"' products with strictly limited functionality (in terms of their ability to manipulate and analyse geographically referenced information). These smaller health care monitoring/planning information systems have typically concentrated on linking together population census information and suitably anonymised geo-coded information on individual mortality or morbidity obtained from annual vital statistics, cancer registers, uniform discharge data sets produced by hospital information systems, and a wide range of other medical registers. In their earliest form, they were little more than automated census-data mapping systems but, more recently, they have become more sophisticated as ICT investment in health care provision has increased, and as progressively larger amounts of health care information have become available in computer-readable and geo-coded form.

The allocation of resources and the planning of provision

In many OECD countries attempts are made to match regional and local levels of health care provision to the level of demand and/or the level of "need". This implies some sort of centralised resource allocation mechanism, be it at the national, regional, or district level. In addition, resource transfer payments between health districts or between individual hospitals are an integral part of many health care financing systems. Clearly, such resource allocation and resource transfer mechanisms are crucially dependent on accurate and up-to-date measurement of supply and demand (and/or "need") and on the ICT systems used to capture that record of supply and demand.

On the supply side, hospital information systems (HISs) provide standard measures of activity levels (numbers of patients, numbers and types of operation, consumable items used and serviced, and so on) whilst the supply of health care via local practitioners is being measured by an ever-increasing array of indicators.

On the demand side, demographic and socio-economic information from the population census (updated regularly to take account of changes in the housing stock, differential birth rates, and so on) can provide expected levels of health care demand for the health district or regional hospital, and geographical variation in that expected demand at the sub-district or sub-regional level can be established using a simple health care planning GIS. Expected demand can then be fed through to establish preventive care targets (*e.g.* for vaccination, immunisation, cervical cancer screening) which local practitioners are required to reach within their "at risk" population. Resource allocations to these local practitioners can then be linked partly to expected demand and partly to the meeting of these preventive medicine targets. Similarly, expected levels of demand can be established for hospitals, and resource allocation can be linked partly to this expected demand, partly to any specialisations provided by the hospitals, and partly (and far more controversially) to the efficiency of provision as measured by RAMOs. Excess demand (or a shortfall in demand) which occurs because of cross-health-district or between-hospital flows of patients can then be compensated by transfer payments.

Planning of new provision – both the level of that provision and the location of any new health care facilities – depends crucially on the same types of supply and demand information as discussed above. In this context, a health care planning GIS is particularly valuable in establishing geographical variation in under-provision or over-provision. Moreover, the linkage of such a GIS to a set of location-allocation computer programmes can be very helpful in guiding the decision on the best location for the provision of any new health care facility (out-patient clinic, hospital, nursing home *etc.*).

Applications of ICT in health care services

Hospital information systems (HISs)

At the heart of current HISs lie the more traditional hospital activity analysis (HAA) and patient administration systems (PASs). These systems were created to: a) record numbers of patients (admissions/discharges), numbers and types of operations, consumable items used and serviced *etc.*, for the hospital as a whole, and b) attach to each patient admitted a unique patient admission/hospital reference number, together with information on the patient's local practi-tioner, the consultant/surgeon who is responsible for the patient, whether the patient has been admitted to the hospital before, the initial diagnosis, the length of stay, the condition of the patient on discharge, and so on. PAS and HAA systems were, and still are, very rarely standardised even within a single region or country. Rather, they are adjusted to local circum-stances and display enormous variability in the way in which they capture information, in the range of information they contain, and in the way in which that information can be accessed.

A sophisticated example of a PAS is provided by Denmark[13]. The Danish system is based upon the land patient register (LPR). This database, which is maintained on behalf of the Danish National Board of Health by Kommunedata, is effectively a shadow copy of the central population register. Individuals receive a national ID number at birth to which basic demo-graphic details (name, address *etc.*) are attached. Each citizen is then legally responsible for providing updated information (*e.g.* within one week of moving house) so that the system can be accurately maintained. What this means, of course, is that most patients' basic details are fully and accurately coded within the PAS prior to any hospital admission and this eliminates the need for time-consuming and error-prone entry of details at the time of admission (the only details collected in this way are on foreigners). In addition, the address information contained within the LPR facilitates simple administration of any transfer payments between health districts (the 14 counties of Denmark) which result from cross-district flows of patients.

Early versions of HISs consisted of separate PAS, HAA, laboratory, medical records, radiology, pharmacy, finance/accounting and personnel modules very loosely connected together. Often the connection was more theoretical than actual. For example, in the mid-1980s leading hospitals such as the Massachusetts General Hospital in Boston and the University Hospital in Tokyo, with very large investments in computer hardware and software, were being described[14] as having a multiplicity of separate departmental systems unable to link to other systems (and occasionally based on separate languages), and as having a non-existent strategy for ICT standardisation, integration and co-ordination.

The key problem of the latter half of the 1980s and early 1990s has been to build integrated HISs in which the separate departmental systems (radiology, laboratory *etc.*) are linked together, and the clinical modules within the system are integrated with the administra-tive/financial modules which support vital functions such as patient registration, ordering of supplies, billing, and the making of statistical returns on throughput, bed utilisation *etc.* The requirement has also been for HISs to be flexible (easily adapted to changes in working practice) and user-friendly.

There are now well-developed examples of successful HISs in virtually every OECD country. For example, the structure of one of the earliest integrated HISs, developed in 1986 for the 1 000-bed Vancouver General Hospital in Canada[15], and linking together a range of Data General, DEC and IBM hardware within a large Ethernet local area network. Similar large-scale systems were also operating at the same time at several other locations, *e.g.* the National Institute of Health Clinical Centre in Washington, D.C.[16]; University Hospital, Leuven, in Belgium; and Leiden University Hospital in the Netherlands. The Leuven system comprised two linked IBM mainframes (devoted to administration modules) and 15 Hewlett-Packard minis for immunology, paediatrics, pharmacology, radiology and pathology. The network was dispersed over 10 kilometres and, together, the mainframes and minis supported 440 terminals in wards and at bedsides. The Leiden system similarly supported 450 terminals, but these were distributed across 13 500 beds in 28 separate hospital units with 1 900 users and it handled more than half a million messages per day. The Leiden HIS was estimated to be costing 1.2 per cent of the hospital's budget and the Leuven system 2.7 per cent[17]. Very large HISs were also being set up at this time in Toronto, Canada, at the Toronto General Hospital, the Western General Hospital and the Wellesley Hospital.

It is of interest to note that the Wellesley was expecting a payback period of only 4.5 years on its investment of $2.2 million (3 per cent of the hospital's annual budget) to develop an HIS supporting 200 terminals[18].

As large hospital information systems became more and more widespread in the late 1980s, user-friendliness and decision-support facilities became increasingly valued attributes. This was particularly the case as nursing sub-systems became integrated into HISs. A typical example of such nursing sub-systems is COMMES, a North American product which was designed to allow the nurse to describe a particular patient care problem to the system and to receive in return a recommendation on possible nursing responses. Following user choice of one or more of these possible responses, the system then derives from its knowledge base specific care plan recommendations to fulfil the needs identified. A similar North American system, HELP, provides diagnostic support by linking the clinical modules in an HIS to a knowledge base of clinical rules. It also prompts the doctor or nurse by alerting them to any potentially life-threatening situation identified as a function of new data being added to the patient's file[19]. Increasingly in the 1990s, other IKBS (Intelligent Knowledge-Based System) "front ends" to HIS will be developed, offering decision support at the bedside on the basis of continuously updated patient records. HISs will, in this way, continue their transformation, from being in many cases merely administrative/managerial tools, to more truly care-oriented systems.

Smart card and network technology

Increasingly, computer-readable patient-held medical record cards are being viewed in many OECD countries as essential elements in an overall health care ICT strategy. The advantages are that the instant availability of medical record information can, by itself, improve the quality of care (particularly in emergency situations), that wasteful duplication of data entry can be avoided, that the quality of patient data can be improved by eliminating opportunities for data entry errors, and that security of medical records can be improved by making each patient the guardian of his or her record.

Patient cards can take many forms – magnetic strip cards, smart cards, memory cards and optical memory cards. Magnetic strip cards have the disadvantage of a small memory size but they can operate with very simple hardware and usually hold sufficient information for easy patient identification. Smart cards, which normally require contact with a computer reader, have larger memory space and greater inherent security, but currently their memory is not large enough for full health care recording needs. Memory cards have the advantage of high memory capacity but do not have the inbuilt intelligence of the smart card. And, in their optical form, whilst having the advantage of unlimited storage, memory cards have the disadvantage of requiring a computer laser to read them – hardware which is both expensive and non-portable. The health care services are currently rather split over which type of card technology to adopt. Many hospitals prefer the unlimited medical record capacity of the optical memory card, whilst local practitioners are supportive of the smart card, or even the simpler technology of the magnetic strip card.

Both large-scale and pilot experiments with patient cards have been undertaken in several OECD countries. In Canada, two provinces have issued magnetic strip health cards to all residents and several of the other provinces are considering smart card pilot projects as a means of health insurance registration and payment. In Spain, a national system is currently being set up which involves issuing 40 million magnetic strip cards to the population. In Japan, memory cards and optical memory cards have been widely used in trial projects involving paediatric and antenatal clinics. In France, the Transvie system uses a smart card as a blood transfusion record.

Typical of the smart card experimental projects at the level of individual urban areas is the Exeter care card project in the United Kingdom. This project, which started in March 1989, involves two hospitals, two local practitioner groups, a dental practice, and every pharmacy in the town of Exmouth. Over 8 000 cards have been issued. Some significant changes have already been observed in the nature of doctor-patient consultations, since doctors are spending less time gathering information and ordering investigations. Reductions in the delays caused by the need to locate patients' records are also producing important administrative economies. Card carrier rates have been measured at more than 80 per cent, and fewer than 1 per cent of those issued with cards have stated a dislike for the system.

A full evaluation of the Exeter experiment is expected by the end of 1990 and it will consider:

- costs (capital, revenue and data loading);
- data-transmission accuracy, reliability, and speed;
- accuracy of the data which are provided verbally by patients in test situations compared with the data held on their cards;
- system reliability;
- attitudes of patients and health care staff;
- card usage rates;
- card security and the risk of information loss if cards are lost or damaged (data must clearly be backed up on something other than a card); and
- major health care effects.

Important challenges for the 1990s concern the adoption of national and international guidelines which will allow patient-held medical records to be made compatible, standards to be set, and standardisation bodies to be established. In this context, the European Commission's Directorate-General XIII (DG XIII) has formed a technical panel to consider European health care ICT issues[20]. Among the issues being considered by the panel are European medical record architectures, definitions for a common European dataset for multi-functional patient cards, the results of European patient data card experiments, and the potential of smart card technologies.

Feeding into the deliberations of the panel in these areas will be the findings of the SESAME advanced-informatics-in-medicine project which is currently addressing communication technologies, knowledge-based applications and data cards.

Finally, it should be noted that smart cards are potentially an important means of access control to networks, and that several European countries are developing health data communications networks which will be of great significance in the 1990s. In the United Kingdom, for example, the Department of Health is currently negotiating a contract for a Family Practitioner Service data network which will link Family Practitioner Committees covering urban areas around the country with the NHS Central Register at Southport. It is envisaged that the FPS network will evolve to become the main NHS data communications network during the 1990s.

Health care GISs

Although the European Commission's DG XIII panel on European health care ICT issues is charged with considering "co-mapping of social, medical and geographical information", and although there has been a very long tradition of epidemiological mapping, plus more recent studies of geographical variations in the use of common surgical procedures[21], the application of GIS technology to health care monitoring and management in urban areas is still in its infancy. To give a flavour of the type of work which has been undertaken so far, four case studies of the use of health care GISs in urban areas in the United Kingdom will be considered. These case studies illustrate two different types of development.

The first type of development involves the purchase by district health authorities (DHAs) and Family Practitioner Committees (FPCs) serving urban areas in the United Kingdom of off-the-peg commercial software (sometimes developed in a totally different context) for use in a health care monitoring and planning context.

Case study N° 1: Hillingdon DHA

A DHA on the western edge of the London conurbation responsible for an urban population of approximately 250 000. In the late 1980s Hillingdon was typical of many DHAs in the United Kingdom, having an unco-ordinated ICT policy which had led to a multiplicity of information systems within the district serving different departmental needs, and with very few links (other than the postcode identifier in the patient records) between the data in these various systems. As a result, health care monitoring/planning tasks with a geographical dimension, *e.g.* the construction of a socio-economic profile for a community nursing neighbourhood, the assessment of small area variations in the uptake rates of screening or immunisation programmes, the presentation of catchment area characteristics for a hospital out-patient department and so on, often presented major problems for DHA staff.

Faced with this situation, Hillingdon chose to experiment with an off-the-peg commercial software package called INSITE which had originally been developed for retailing and market research[22]. This software facilitates the linkage of postcode-referenced health records to small area census information via a postcode-to-census-area directory. Simple geographical manipulation of the health records and census information is then possible. For example, the database can be queried on the basis of either postal area or census area geography, or for ad hoc areas such as circles and segments. In the latter case, co-ordinates of the polygons which define the ad hoc area are entered, and summary statistics on socio-economic characteristics, disease incidence, uptake of immunisation or screening, and so on, are calculated for the area. The computation performed by this software is less explicitly spatial than in a true GIS, and much of its power rests on the accuracy of the postcode-to-census-area directory. In addition, unlike a true GIS, it has no integrated graphics capability. However, it produces relatively cheap and simple solutions to some of the routine health care monitoring/planning tasks which health authorities in UK cities face. INSITE can be linked to a simple off-the-peg mapping package such as ATLAS to provide the cartographic representation often required in reports of the district's public health, and in public presentations.

Case study N° 2: Barnsley FPC

Family Practitioner Committees in the United Kingdom are the administrative organisations responsible for the regulation and payment of local practitioners (GPs), dentists, and other health care professionals providing primary medical care through the mechanisms of the NHS. More than half the administrative staff in a typical FPC are concerned with maintenance of the patient list which contains basic information about every individual registered with a GP, plus cancer-screening results. The list is constantly updated as patients move and re-register with GPs. Barnsley FPC, which covers an industrial urban area in the north of England, like many FPCs in the United Kingdom, has traditionally had little ICT investment and little capacity to utilise the information "locked up" in its expensively maintained patient list. Recently, however, together with other FPCs, it has acquired a commercial PC-based mapping system, PC-Mapics, in an attempt to make greater use of its patient list information for health care planning purposes[23]. Using this system, age/sex/occupation breakdowns of the patients associated with each GP can be mapped, offering the potential to create socio-economic profiles of practices and assessment of the "market penetration" of each GP practice. This type of information can then be used to assist in the redistribution of patients between new surgeries in the urban area when an existing practice has been split, and in the re-allocation of patients between GPs when they are too far away from the surgery to be cost-effectively served by their GP. In addition, the type of information contained within the FPC mapping system will be essential to both GPs themselves and the FPC and DHA in establishing the basic patient demand information which will be required as the NHS in the United Kingdom moves towards a more market-based financing system (see below).

The second type of development involves the investment made by certain DHAs in building customised health care planning GISs which are not constrained by the limitations of off-the-peg commercial software.

Case study N° 3: the Avon health care planning system

In 1985-1986 the three DHAs covering the city of Bristol and the county of Avon in south-west England (with an urban population of 800 000) sponsored the development of a prototype health care planning GIS. The lead DHA, Southmead, in particular, faced a number of problems which were essentially geographical in nature – claims that certain industrial areas of the city of Bristol suffered from unusually high mortality and morbidity, several small areas of relatively high social deprivation, and a mismatch between the geographical distribution of population and of hospital facilities in the DHA. Southmead DHA was unable to address these issues via its existing systems, and off-the peg software such as that employed by Hillingdon and Barnsley was not as readily available in 1985-86 as in the late 1980s. A group of university academics was therefore commissioned to create a prototype system[24].

This involved the creation of a digital map base for the county of Avon at the census enumeration district level; the linkage of census information and postcoded mortality and cancer-registry data into the digital map base; and the development of a simple IBM-PC-based mapping, display and query system known as Map Manager.

Using this system, DHA personnel could then display, and "browse" through in an interactive fashion, maps of overlaid census and geo-coded health information, they could "window" into and enlarge very small areas of the DHA (see Figures 5 and 6), and they could answer specific queries about the geography of demand for health care in the district. The system could also be used to map the supply of health care facilities, to assess the interaction of supply and demand, and to help plan the location of new facilities. However, in its prototype form it contained only very limited analytical functions. Rather, it served as a low-cost entry point for the DHA into the technological possibilities of large-scale commercial GISs, and as a starting point for more detailed GIS-type investigations of such issues as the provision of GP family planning services.

Case study N° 4: a health care GIS for the city of Sheffield

In 1989, Sheffield DHA began a joint project with the University of Sheffield which essentially built on the Avon prototype and also on earlier work undertaken by the DHA concerned with a system of health "profiles" for the electoral wards within the city. Sheffield DHA is a major contributor to the World Health Organisation's "Health for all by the year 2000" campaign, and it holds very large quantities of postcoded information, including records of all hospital admissions, birth and deaths, attendances for vaccinations and cervical

Figure 5. **Avon district health authorities -- index map**

Index map of Avon health authorities as it appears on PC screen to health managers
at beginning of an interactive session.
Shows two areas "windowed" into for further analysis.

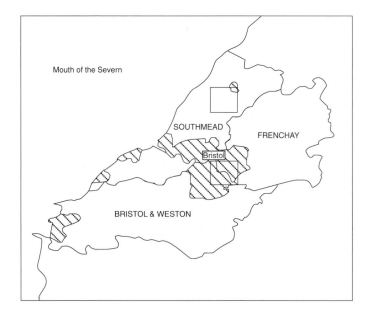

Figure 6. **Avon district health authorities -- enlarged map**

Enlargement of a small part of one of the "windowed" areas,
showing census enumeration district boundaries, and location of
cases of a particular type of disease within a radius of 450 metres
from a particular location (*e.g.* an industrial plant).

smear tests *etc.* Postcoded information also exists relating to community nursing and child
health services, and out-patient information will become available in 1991. The aim of the joint
project is to construct an integrated health care GIS which can exploit this rich data resource,
can examine variations in mortality and morbidity within the city, and offer a greater range of
analytical functions than the prototype Avon system.

The project has four distinct stages[25]. The first is the construction of a common database,
involving the integration of the postcoded information from the various operational systems
cited above. Census small-area statistics from both the 1981 and 1991 censuses (when available) will also be added, together with information from other databases, such as the topographic
information currently held by the City Council. The second stage will involve the development
of an areal framework for the analysis of morbidity and mortality data. A fundamental difficulty with the use of vector-based GISs for population-related applications is that the areal units
used (*e.g.* census zones) are arbitrarily imposed on the underlying socio-economic patterns.
This problem is well known in academic geography[26], but has been insufficiently addressed in
the context of GISs. The inclusion of research into these aspects is an important component of
the project. The third part of the work will be the development and adaptation of spatial

171

statistical methods to standardise and evaluate morbidity and mortality patterns, and the final aspect of the project will be to construct an operational system for monitoring spatial variations, which can be used by a wide range of personnel in the DHA.

The Sheffield project represents a very careful application of ICT to a DHA's spatial information needs. Research into the implications of using a variety of areal bases and the effects of different spatial statistical techniques will form a major part of the work and the resulting system will incorporate the findings of this analytical work. This is in contrast to the direct implementation of a commercially available GIS solution with an existing database. When information requirements are immediate and software is in use for similar problems elsewhere, it is tempting to attempt the transfer of technology from one situation to another, without detailed consideration of the unique nature of the user's data. The Sheffield project promises to produce an integrated system which is optimally suited to the nature of the databases involved, and will be of value to the DHA, to the City Council, and to researchers whose interests are more theoretical.

The institutional context of expanding ICT provision in health care services: a case study

It was noted that the wide variations in the types of health care system operated by OECD member states strongly condition the nature of ICT take-up and usage. Also, it was noted that increased levels of ICT expenditure on health care are intimately connected to the search for efficiency of provision in an era of restricted budgets, and to transformations of existing health care financing and delivery systems into more market-based systems of "managed competition". Nowhere is this seen more clearly than in the recent UK Government's proposals to transform the nature of the UK NHS. The 1990 NHS and Community Care Bill, and the White Papers "Working for Patients" and "Caring for People" which set out the concepts underlying the NHS Bill, propose radical changes to the structure of the NHS. Most importantly in the context of this Handbook, these proposals have very significant implications for ICT in UK health care provision. Stated very simply, the provisions of the NHS Bill attempt to set up an internal market within the NHS. Providers of acute care services (hospitals *etc.*) will on 1st April 1991 become dependent for their income on attracting contracts for the services they provide. These contracts will come from the purchasers of health care services – District Health Authorities (DHAs) or groups of local practitioners (GPs) who are assigned funds to hold and manage (GPFHs) on behalf of their patients. The belief is that DHAs and GPFHs, by exercising choice, will create competitive pressures and, by specifying service quality levels, will improve value for money. The services to be provided will be determined by DHAs' and GPFHs' appreciation of the health needs of the population they serve.

In this new environment, all parties to the process have stressed the crucial supportive role which ICT systems will play, and a truly significant amount of effort and new money is currently being pumped into developing both immediate ICT responses to the proposals and a longer-term ICT strategy for the new market-based NHS[27].

For many years, the NHS has been described as being "data-rich, information-poor". But in the post-NHS Bill environment, accurate information and the ability to access and utilise that information will be the key to the successful operation of the internal market from the point of view of both the purchasers of services and the providers. The purchasers (primarily the DHAs through block contracts), for example, must be in a position to negotiate with the providers. This implies that:

- they must know the characteristics and health status of the population resident within their districts;
- they can establish a method of assessing the future health needs of the resident population;
- they have accurate information on the current patterns of patient referral to acute providers by GPs within their districts;
- they can keep track of interactions by all residents of their districts with the various parts of the health care system;
- they can oversee the requirements of residents awaiting treatment and minimise waiting lists by contracting with other providers.

Clearly this is a formidable list and it implies some significant changes in the information collection and management operations of DHAs.

At present, DHAs are hospital-oriented in their information collection and management activities. They obtain information on the patients treated by the hospitals which lie within their boundaries, *i.e.* on the catchment population of the hospitals, not on the population resident within the area of the DHA. Therefore, the DHA does not get to know directly about treatments which are received by its residents outside its boundaries. Following the change in the NHS system, the information collection and management orientation of a DHA must change to be based on its resident population – for they are the patients for whom it will be billed by the providers. This implies that each DHA will have to develop an up-to-date register of its population (names, addresses, sex, ages, NHS numbers *etc.*). Also, in order to understand the characteristics of its population, plus the current health status and future health needs of that population, the DHA will need to bring together, at census district level, information on social deprivation, ethnic origin, age structure, interactions of residents with health care services, immunisation status, and so on. In other words, DHAs can be expected to move very rapidly towards developing the type of databases which underlie the health care GIS applications discussed above. Given the addition of interactive graphics and simple spatial analysis functions to these databases, it is clear that the 1990s will see health care GISs becoming the major ICT tool of the demand/purchaser side of the internal health care market in the United Kingdom. And alongside this must come: significant improvements in the coverage and accuracy of geo-coded (usually postcoded) health records, the establishment of the "minimum data sets" required for the working of the new contracts, the linkage of the DHAs into the Family Practitioner Service data network and thus to the NHS Central Register (see above), the development of information communication standards and the preservation of the security of patient information within networks.

From the point of view of the providers (hospitals *etc.*), the post-NHS-Bill internal market will require them to set prices for their services, and to offer GPFHs prices for (and invoices for) each case of in-patient care, each out-patient visit, and each GP direct referral for diagnostic tests *etc.* In addition, they will have to negotiate block contracts with DHAs, and provide regular statements of the number of completed (and uncompleted) episodes of care under the contract within given periods of time. That implies that they must be able to:
- associate each patient with a "payer";
- accurately characterise every service rendered;
- provide any other information (on length of stay, waiting time *etc.*) which the payer may require.

173

To identify the payer will require accurate data on the address (and postcode) of each patient, the GP, country of residence if outside the United Kingdom *etc.,* plus accurate and continuously updated computer-readable national directories on: address to postcode; postcode to district of residence/DHA; GP to GPFH; county of residence to reciprocal health agreement; and so on. In addition, some "payers/purchasers" (*e.g.* the DHAs) might have more than one contract with a particular provider and thus each patient will have to be associated with a particular contract. This, in turn, is likely to require information on diagnosis, operation type, speciality, particular types of test undertaken, clinic type, consultant name, and so on. Although these requirements are simple to state, they nevertheless impose data collection and information handling targets which typically lie outside the capabilities of the PASs and HISs currently found in UK hospitals. And they lie well outside the current capabilities of the providers of non-acute services (community health services, long-stay hospitals *etc.*) who typically have only recently taken the first steps in ICT installation. It is clear, therefore, that major ICT investment will be required and that lack of skilled personnel will present a major problem.

ICT in UK health care provision is currently being discussed more widely and more effectively than at any time in the past. Indeed a national health care ICT infrastructure is in the process of being created, and the UK Government is providing significant new resources to facilitate this development. The purpose of this case study, however, has been to emphasise that these recent forces of innovation in ICT take-up and usage in UK health care have not been led by clinical considerations or by any radical developments in technology itself. The institutional context of expanding ICT provision and usage is political. ICT is seen as a means by which a market-based system of health care provision in the United Kingdom can be put into operation. Moreover, any consequences for the detailed operations of local urban government of what are fairly dramatic changes in the way in which health care information will be collected and managed in the United Kingdom will, to a large extent, be unintended.

Assessment

The urban context of health care ICT

Unlike many of the functions dealt with in this book, health care delivery in many OECD countries does not form part of the direct responsibilities of urban local government. In some countries, such as the United Kingdom, a completely separate hierarchy of administrative areas (in the United Kingdom these are referred to as health regions and health districts) has been created to provide health care services, and health authority boundaries often cut across both local government boundaries and the physical extent of urban built-up areas. Figure 7, for example, shows the three health districts, Bristol and Weston, Frenchay, and Southmead (parts of the South Western Regional Health Authority) which serve the city of Bristol and most of the county of Avon. It will be noted that:

– these health district areas are not totally coincident with the boundaries of the county (which is the upper tier of local government in the United Kingdom) – the city of Bath which falls within Avon is served by another health district which, confusingly, is not even part of the South Western RHA; and
– the health districts sub-divide the city of Bristol, thus cutting across the boundaries of the lower tier of local government.

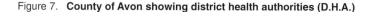

Figure 7. **County of Avon showing district health authorities (D.H.A.)**

In other OECD countries, such as Sweden, whilst health care is organised nominally within local government, in practice it is almost as separate a system as in the United Kingdom. In Sweden health care is provided by the County Councils (which include within their number some of the larger independent cities). However, health care is practically the only responsibility of the County Councils, accounting for 85-90 per cent of their operating costs.

Elsewhere, a rather more confused pattern of responsibility exists. In Germany, for example, hospitals are owned by states, counties, city governments and churches. Although funding comes from state level, there has traditionally been great variation between the 11 states which composed West Germany, and it is widely acknowledged that there are very few communication links between state health authorities, hospitals, GPs, social/medical insurance companies, and city government.

Finally, it has already been noted that health care delivery in the United States is decentralised, pluralistic, and dominated by the private sector. Local government is effectively involved only in the delivery of health care to the 35 million Americans who have no financial protection against medical expenses, and this "last resort" provision clearly has some very special characteristics.

It is against this background of a wide spectrum of health care systems, and what is often a very tangential relationship between the responsibilities of urban government and the delivery of health care to the urban population of OECD countries, that this chapter must attempt to assess the impact of health care ICT on urban functioning. What then, if anything, can realistically be said? And are there any policy options that have more than strictly local significance?

Policy options for an era of market-based "managed competition"

Given the considerations outlined above, the only realistic way, in the author's opinion, to suggest policy options relating to the impact of health care ICT on urban functioning is to take as a starting point the almost universal movement in OECD countries towards more market-based systems of state-financed public health care delivery – systems which are characterised by what is termed managed competition. In these systems, as illustrated by the changes currently taking place in the UK NHS, and as seen strongly elsewhere in Europe, ICT has a vital role to play in putting the market into operation. Without significant ICT investment, without effective management of information which has traditionally been locked up in health providers' files, without new types of database, and without new methods of manipulating and displaying these databases, it will prove extremely difficult for the purchasers of health care services (be they health authorities, counties, states, city governments or insurance companies) to assess the levels of current and future demand for particular types of health care services by the populations of particular parts of the country (regions, counties, urban areas, or whatever). Similarly, without significant ICT investment and the development of integrated HISs, it will be extremely difficult for the providers of health care services (hospitals *etc.*) to establish the prices which are required for negotiation of contracts between themselves and the purchasers, and for the operation of the market. Yet many OECD countries are placing considerable faith in the development of a quasi-market (of managed competition) to create the competitive pressures which they see as being vital to the improvement in the value-for-money obtained from their very significant public health care expenditure.

What then are some of the policy options which are open to urban governments within the context of these emerging, heavily ICT-dependent, market-based systems of health care provision?

Joint contribution to the development of integrated urban databases

Purchasers of health care services require up-to-date registers of population, integrated at a census district level with information on matters such as social deprivation, demographic structure, ethnic groupings, disease incidence, immunisation and screening, the interactions of residents with local practitioners and acute services, and so on. Urban governments often require very similar information (not least to collect local service provision taxes), together with information on the physical structure of the urban built form: roads, railways, utility networks, industrial land use, public open space, housing structures, and so on. Currently, there is very little interaction between health care databases and urban management databases in OECD countries, and much wasteful duplication of effort and unnecessary expenditure. Yet the potential for worthwhile co-operation between the database managers of the health care purchasers and their counterparts in urban government is enormous. Moreover, recent ICT developments have ensured that the problems of confidentiality of information and security implicit in such co-ordinated databases are relatively minor. Controlled access to databases through IKBS "rules-based" front ends has been shown to be possible, even for the largest national databases[28], and the 1990s will undoubtedly see considerable development of this technique.

Shared investment in GISs by both urban government and "health care purchasers"

The prototype studies discussed above have indicated some of the potential which health care GISs have for the purchasers of health care services operating in the new market-based systems of provision. Urban governments in OECD countries have also experimented with GISs, particularly in the fields of land use and transport. However, the driving forces behind the development of GISs in European OECD states have been the utility companies (water, electricity, gas, telecommunications) which have made multi-billion-dollar investments in appropriate hardware and software. Given the development of, or at least the movement towards, integrated urban databases, shared investment in GIS capability between an urban government and the health care purchaser who covers that particular urban area would seem to be a logical way forward – particularly as this is a high-cost field in which to develop an effective capability, and one in which trained personnel are in very short supply. In this way, the essential compatibility of technology will also be ensured, and the costs of training the necessary skilled personnel can be shared.

Demonstration projects

If policy options 1 and 2 are felt to be far too radical to contemplate on any national scale, then there would be considerable value in considering a very small number of demonstration projects, backed perhaps by some cross-national funding. As noted before, currently the great majority of ICT investment in health care goes into the provider side of the system and it was the success and cost-effectiveness of the pilot versions of the integrated HIS which spurred this pattern of investment. Successful ICT demonstration projects, linking integrated urban databases and health care GISs, on the demand/purchaser side of the system, are required if this pattern of ICT investment in health care provision is to be radically changed.

Shared smart card experiments

The patient card experiments discussed above can quite clearly be extended to include services whose provision typically falls within the responsibilities of urban government. The exclusion of social/welfare services in an urban patient card experiment such as that being conducted in Exeter, which includes a large health centre, does indeed seem extremely arbitrary. What is required is what in the market research literature is referred to as a "test town", in which all medical, social and welfare facilities are integrated into a single smart card experiment.

Shared networks

Smart cards also offer access control to networks. Clearly, as health data communications networks evolve in the 1990s, urban government will find it increasingly attractive and beneficial to have a certain degree of access to these networks. Once again, security and confidentiality of information issues will be prominent. So, smart-card access control, together with IKBS rules-based access control, must feature strongly in any moves towards shared networks.

A more radical integration of the responsibilities of health care purchasers and urban government?

The structure of health care delivery in many OECD countries has developed in isolation from the structure of local government and has traditionally been provider-oriented, that is to say, the management of the health care system has been structured around acute provision facilities. The movement towards market-based systems of health care provision implies, however, that the purchaser side of the system is separated out in a very clear manner, and it gives that part of the system a set of well-defined responsibilities. The more these responsibilities are considered and set alongside the service provision responsibilities which have tradition-ally fallen within the ambit of local government, the more obvious it becomes that there may be scope for a much closer integration of block-contract health purchasers with local government. Certainly the type of mismatch between health care provision districts and local authority areas which has arisen in certain OECD countries because of a hospital-centred division of the urban population is totally illogical in a market-based system where purchasers can contract with any number of providers and are not tied to supporting a particular acute care facility. Given the information needs of the purchasers in market-based systems, and the congruence between many of the ICT requirements of such purchasers and those of local/urban government provid-ers of closely-related services, how long can it be before there is a breakdown of the sort of health care purchaser areas and a reorganisation of them into a closer relationship with local authority/city government areas? Or will inertia prevail, and the separation of purchaser and provider so necessary to the operation of a public health market-based system prove to be more illusory than real?

NOTES AND REFERENCES

1. SCHIEBER, G.J. and POULLIER, J.P. (1990), "Overview of International Comparisons of Health Care Expenditures" in *Health Care Systems in Transition: The Search for Efficiency,* OECD Social Policy Studies N° 7, OECD, Paris.

2. (1990), "Minister Launches IT Strategy", *British Journal of Healthcare Computing,* 7(1), 9 February.

3. ENTHOVEN, A.C. (1990), "What Can Europeans Learn from Americans?" in *Health Care Systems in Transition: The Search for Efficiency,* OECD Social Policy Studies N° 7, OECD, Paris.

4. HIMMELSTEIN, D.U. and WOOLHANDLER, S. (1986), "Cost Without Benefit: Administrative Waste in US Health Care", *New England Journal of Medicine,* 314(7), pp. 441-45.

5. LEKIE, D. (1987), "USA Health IT", *British Journal of Healthcare Computing,* 4(2), pp. 30-32.

6. ENTHOVEN, A.C. (1990), *op. cit.*

7. Information taken from "National Contribution from France" drafted by URBA 2000, to OECD Project on Urban Impacts of Technological and Socio-Demographic Change. Theme B: Impacts of Information and Communication Technologies on Urban Functioning. OECD, Paris. Paper OP/TSDB(89)8.

8. WILLIAMS, R., CUNNINGHAM, G., NORRIS, F. and TASHIRO, M. (1976), "Monitoring Perinatal Mortality Rates: California 1970 to 1976". *American Journal of Obstetrics and Gynaecology,* 136, pp. 559-568. Also STEINBROOK, R. (1987), "Care for Newborns Varies, Studies of Hospitals Show" *Los Angeles Times,* 9 November.

9. STEINBROOK, R. (1988), "Heart Surgery Death Rates Found High in 1 in 6 Hospitals", *Los Angeles Times,* Part 1, 24 July.

10. STEINBROOK, R. and ROSENBLATT, R. (1987), "US Issues Data About Hospitals' Death Rates", *Los Angeles Times,* 18 December.

11. EVANS, R.G. and BARER, M.L. (1990), See "Respondents' Comments", in *Health Care Systems in Transition: The Search for Efficiency,* OECD Social Policy Studies N° 7, OECD, Paris.

12. DEPT. OF PUBLIC HEALTH MEDICINE, SOUTHMEAD HEALTH AUTHORITY (1988), *A Report on the Health of the District's Residents,* Bristol (typical example of such public health report).

13. FOSTER, I. (1989), "Info Systems: The Danish Example", *British Journal of Healthcare Computing,* 6(9), pp. 36-39.

14. LECKIE, D. (1987), *op. cit.,* p. 32 and PEEL, V. "IT for All by the Year 2000?", *British Journal of Healthcare Computing,* 4(6), p. 32.

15. WINDSOR, P. (1986), "Computers in Canada's Hospitals", *British Journal of Healthcare Computing,* 3(5), pp. 19-21.

16. LECKIE, D. (1987), *op. cit.*

17. WINDSOR, P. (1987), "Worldwide Healthcare Computing", *British Journal of Healthcare Computing,* 4(1), pp. 14-19.

18. WINDSOR, P. (1986), *op. cit.*

19. WINDSOR, P. (1987), *op. cit.,* p. 19.

20. GUINNANE, J. (1989), "The Health of Nations", *British Journal of Healthcare Computing,* 6(9), pp. 23-25.

21. MCPHERSON, K., WENNBERG, J., HOVIND, O. *et al.* (1982), "Small-area Variations in the Use of Common Surgical Procedures: An International Comparison of New England and Norway", *New England Journal of Medicine,* 307, 1310-1314. Also MCPHERSON, K., STRONG, P.M., EPSTEIN, A. and JONES, L. (1981), "Regional Variations in the Use of Common Surgical Procedures", *Social Science in Medicine,* 15A, pp. 273-288.

22. BLANKLEY, N. (1990), "Local Integration of Health Care and Geographic Information", paper presented at Geographical Information Systems and Health Care Analysis Conference, organised by the Operational Research Society, June, London.

23. MIDGELEY, J. (1990), "FPCs and Mapping", paper presented at Geographical Information Systems and Health Care Analysis Conference, organised by the Operational Research Society, June, London.

24. WRIGLEY, N., MORGAN, K. and MARTIN, D. (1988), "Geographical Information Systems and Health Care: The Avon Project", in *Working with Geographical Information Systems,* ESRC Newsletter 63, pp. 8-11.

25. DISTRICT HEALTH AUTHORITY OF SHEFFIELD (DHA) or HAINING R., Department of Geography, University of Sheffield, S10 2TN, United Kingdom.

26. OPENSHAW, S. (1984), *The Modifiable Areal Unit Problem,* Concepts and Techniques in Modern Geography N° 38, Geo Books, Norwich, United Kingdom.

27. DEPT. OF HEALTH, HMSO, (1990), *Framework for Information Systems: An Overview,* Working Paper N° 11, London.

28. RHIND, D. (1985), "Successors to the Census of Population", *Journal of Economic and Social Measurement,* 13, 29-38. Also RHIND, D. (1987), *Creation of a Prototype On-line IKBS for Handling Census Data,* ESRC Research Award N° HOO232091, Economic and Social Research Council, London.

CASE STUDIES OF ICT USE IN HEALTH CARE PROVISION IN URBAN AREAS

Case study N° 1 – A health care GIS for the city of Sheffield

Contact persons:

– Medical   Dr. Dorothy Birks
                    Consultant of Public Health Medicine
                    Sheffield Health Authority
                    Westbrook House
                    Sharrow Vale Road
                    Sheffield, S11 8EU, United Kingdom
                    Tel: (0742) 670333, ext. 6186
                    Fax: (0742) 660498

– Academic   Dr. R. Haining
                    Reader in Geography
                    Department of Geography
                    University of Sheffield
                    Sheffield, S10 2TN, United Kingdom
                    Tel: (0742) 768555
                    Fax: (0742) 739826

Brief summary

An integrated health care GIS currently in the process of being developed for the city of Sheffield, United Kingdom. A joint project between Sheffield District Health Authority and the University of Sheffield which has the aim of developing an operational system to be used for monitoring, assessing, and managing variations in mortality and morbidity within the city. The system aims to link together small-area statistics from the UK 1981 and 1991 population censuses with geo-coded information on births, deaths, hospital admissions, attendance for vaccinations and cervical smear tests, out-patient clinics, community nursing and child health. In addition, other data held by the City Council, *e.g.* topographic information, will be linked into the system. A programme of development and adaptation of spatial statistical methods is also to be undertaken to provide appropriate analytical techniques for use within the system.

Case study N° 2 – A smart card system of patient-held medical records: the Exeter care card project

Contact person:

Dr. Robin J. Hopkins
Department of General Practice
Postgraduate Medical School
University of Exeter
Barrack Road
Exeter, EX2 5DW, United Kingdom
Tel: (0395) 272979
Fax: (0395) 276156

Brief summary

A pilot project to investigate the potential of patient-held medical records on smart cards within a single urban area. Project conducted in town of Exmouth, United Kingdom (population 32 000) during 1989/90, and funded by the UK Government's Department of Health. Some 8 500 smart cards were distributed and the trial involved 2 hospitals, 2 local practitioner (GP) groups, a dental practice, and 8 pharmacies within the urban area. Card carrier rates measured at over 80 per cent and very few patients stated a dislike for the system. Some significant changes observed in nature of doctor-patient consultations and in number of general clinical investigation ordered by GPs, together with improvements in accuracy of data and other benefits.

NEW TECHNOLOGIES IN PUBLIC TRANSPORTATION AND TRAFFIC MANAGEMENT

Jürg SPARMANN
Studiengesellschaft Nahverkehr GmbH
GERMANY

Jürg Sparmann received his master's degree in civil engineering in 1970 from the Technical University of Berlin, where he then worked as a research and teaching fellow. In 1978, he received his doctorate for his thesis in the field of traffic engineering and control. After two years at the University of California-Berkeley, where he received his Ph.D in the field of transportation economics, he joined SNV, a study group for urban transport, research and development and became general manager of the Berlin branch of SNV.

He is project manager of the LISB field trial and has been appointed as a member of the technical evaluation and follow-up Commission of the DRIVE programme of the European Economic Community.

INTRODUCTION

Innovative information and communications technologies have contributed to efficient traffic management in the past and, as they develop, are expected to bring substantial improvements in the future.

However, the extent of the improvements will depend on the constraints under which the systems operate. Tremendous changes in traffic patterns have occurred; the growth of the motor car has exceeded expectations in the 1970s. In 1970, it was estimated that there would be 30 million private cars in the Federal Republic of Germany in the year 2000, but this figure was reached in 1990. Revised estimates suggest a further increase of at least 20 per cent. This means that about 20 per cent more parking spaces will be needed and that traffic volume will increase by about 12 to 15 per cent. Similar trends are forecast in all other OECD countries (Figures 8 and 9).

This will place a tremendous burden on our environment, particularly in urban areas. It is no longer possible, in most cases, to increase road capacity by extending cities' road networks. The same is true of car parking space. These problems must be solved by the end of the century if a complete traffic breakdown is to be avoided. Therefore public transport will become even more important.

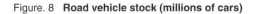

Figure. 8 **Road vehicle stock (millions of cars)**

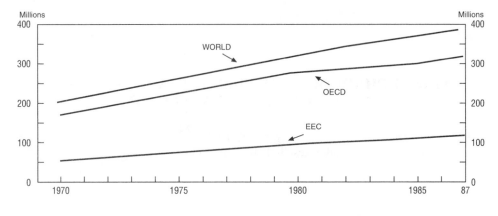

Source: OECD.

Although public transport operators have put a great deal of effort into modernisation, in order to offer good service at a reasonable price, their passenger numbers have decreased. They should not be discouraged by this; without the improvements achieved so far the situation today would be even worse. Unfortunately, they have also been forced to become more cost-effective because of reduced public subsidies. In view of the social costs created by private road traffic, these subsidies can be seen as compensation for the disadvantages suffered by public transport in a competitive transport market.

Figure 9. **Road traffic volumes (billions of vehicle-km)**

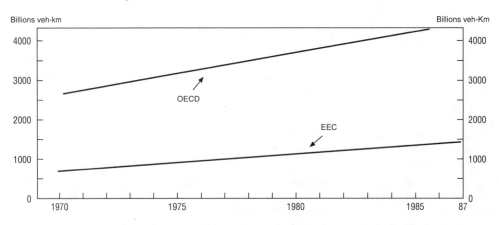

Source: OECD.

184

However, if public transport is to become a more acceptable alternative to the private car – and this has to be one of the objectives of future traffic management strategies – its quality of service must be improved considerably.

Increased use of the motor car is not responsible for all the problems of our cities. In recent years people have demanded a better quality of life; this includes the protection of the environment from noise, pollution and other nuisances caused by urban traffic. However, access to our cities has to be guaranteed to avoid restricting economic growth; moreover, personal travel patterns have become more flexible, and most people believe their needs can only be satisfied by using private cars. This argument, based on people's perceptions, is one of the sources of the present conflict: each individual tries to optimise his/her travel pattern regardless of the impact on society.

There is hardly a day or a route on which traffic congestion can now be avoided. Congestion causes increased fuel consumption, exhaust emissions, travel times and social costs, and last but not least affects safety – and all of these impose a burden on a city's economy and community. Much effort has been put into improving traffic management systems, but even highly sophisticated systems have their limits; this is also true of the new technologies being developed and tested in this field. Therefore, it is urgent to study the pattern of people's door-to-door movements to seek a solution.

In the search for a solution, the attitude of people to their transport needs, and their behaviour in choosing different modes of transport – private car, public transport, bicycle or walking – are important. If it is true that attitudes are changing, it is vital to harness this change to influence behaviour and choice.

Traffic management in the 1990s has to be planned on the basis of the present road network. It follows that a satisfactory solution to future traffic problems can only be achieved if a further increase in traffic volume is avoided, at least in those areas where road capacity has already been reached. Traffic management systems can work only if the road network has not reached saturation point. Even if they help use the roads more efficiently, they cannot increase overall capacity.

Since the limit has been reached, particularly in city centres, people have to be convinced that they will not in future be able to use their private cars to go anywhere, any time. We shall be forced to impose more restrictions on access to city centres by private cars while offering alternative transport in order to keep the city centres alive. Providing public transport which treats passengers as customers rather than as ciphers will become a greater challenge in the future than it is now. At the same time, people will have to learn to be more selective in using their cars; imposing restrictions is the last resort, but it becomes necessary if no other solution is satisfactory. Restrictions can be imposed by regulations or by charging drivers for using roads and parking spaces, particularly in overloaded and ecologically sensitive areas, either all day or at certain times.

State of the art and strategies

For the sake of clarity, a brief definition should be given of the objectives and scope of dynamic traffic management. Traffic management covers all forms of action to influence road use, and includes parking. Dynamic operation includes action in response to accidents, changes in weather *etc.,* as well as traffic control by means of control infrastructure, *e.g.* traffic signals, variable message signs, queue-warning systems, reversible-lane operation and parking-guidance systems.

The traditional objective of traffic management is to optimise traffic flow, subject to the forces of traffic demand and road infrastructure supply. Putting it another way, traffic management systems help meet travel needs generated by the economic and social activities of people. Constraints on traffic are imposed by resources, *i.e.* road capacity, parking availability, environmental impact, safety, energy use, and other social, economic and ecological concerns.

However, in many European cities a much broader view is emerging, as existing traffic management techniques reach the limits of their potential. This view recognises that the road infrastructure is used by different modes of transport (*e.g.* public transport, business traffic, goods transport, private traffic) which have to be accorded different priorities, depending on the time of the day and the area involved. It sees that increasing mobility and traffic volume will necessitate shifting a substantial part of travel demand from private cars to public transport, in order to use transport capacity more efficiently – provided that the public transport alternative is acceptable. For this reason, public transport must become significantly more attractive and more competitive with the private car. This has led many cities to introduce special bus lanes which can also be used by taxicabs, and communication systems between vehicles and roadside equipment to give priority to buses and light rail vehicles by prompting a green light when these vehicles approach traffic lights. Important as these measures are, they are only a small first step towards integrating public transport into an overall dynamic traffic management system.

Freight and commercial traffic are also important to the functioning of cities. Traffic growth creates difficulties for delivery drivers. In many instances they have to double-park, obstructing other road users, because there is no parking space available for loading and unloading. They also get trapped in traffic jams. As a result, delivery of goods becomes more and more unreliable and customers become more and more dissatisfied. This means operators of delivery fleets are forced to use extra vehicles to ensure reliable service; as a consequence operating costs go up, as do traffic volume and environmental impact.

Therefore, traffic management strategies have to take particular care to improve delivery services, including better route planning and fleet control, and much closer co-operation between fleet operators. Overall vehicle mileage could be substantially reduced if the delivery of goods were organised in such a way that, for instance, food shops selling different kinds of bread received supplies in one vehicle instead of different vehicles for the different types of bread. Freight distribution centres should also be considered as an important step towards the more efficient and environment-friendly distribution of goods.

Last but not least, traffic management has to make private traffic more economical and environment-friendly. With public transport the most important consideration is to run to schedule, but private road users expect other benefits from traffic management – time savings by avoiding congestion, cost reductions, and a high degree of safety. These personal objectives are in many instances out of line with the objectives of the community as a whole. While the private car user expects individual travel time to be minimised, the community wants total travel time reduced, and subjected to environmental and ecological constraints. Overall, the most important criteria for traffic management are safety and environmental protection. The traffic management strategy has to be a compromise giving everybody as much satisfaction as possible, in order to achieve a high acceptance level for control measures.

It also has to provide as much flexibility as is needed to enforce transport policy objectives; this means using different methods depending on the time of day (peak or off-peak hours) and the area (*e.g.* town centre, residential area).

The integrated approach proposed for future traffic management is based on overall supply and demand for the transport infrastructure, including public transport systems. This is particularly important because some forms of public transport use the road infrastructure; and there is a relationship between the quality of public transport and traffic flow on the roads, which affects how people choose to travel. The integrated approach becomes more important when the road capacity is reached and further extension of the road network is neither possible nor ecologically desirable.

Demand is not an end in itself, but a consequence of travel needs caused by activity patterns. However, the activity patterns of people are the result of the distribution of activities within an urban environment. Therefore, a city's demand structure depends on its land use structure. Given its land use, demand can only be influenced marginally unless mobility is restricted. However, traffic management systems can prompt changes in how people choose to travel, by providing better information for them to plan their trips (*e.g.* selecting the best time for travelling and considering alternative destinations if possible) and by providing demand-responsive information for route selection whether travelling by car or by public transport. In this respect, user information systems will become more and more important in future. New developments in information and communication technologies can help solve this problem.

On the supply side, traffic management systems can affect the standard of service and help use road capacity more efficiently. As far as public transport is concerned, only a high level of service will increase acceptance and influence people to choose it. However, the traffic problems of the future cannot be solved without considering restrictions on the use of private cars when other solutions alone do not succeed. As these restrictions are politically undesirable, even more effort has to be put into improving public transport so that people use it voluntarily.

Better use of the road infrastructure can be achieved by means of traffic signals, road signs, traffic messages and restrictions, but tolls and road pricing also have to be considered as future options. As there is a close relationship between public transport and the management of road traffic, future efforts should be directed towards demand management. By providing adequate information, choice as to mode of travel, time of departure, optimal routing, and even the decision to travel or not, can all be influenced.

It is of the utmost importance that management should be as dynamic (traffic-responsive) as possible. Users will accept a change in their normal travel patterns if they feel they can trust the traffic information they are given. Therefore, the credibility of traffic information is fundamental, if future traffic management systems are to operate satisfactorily. This means looking for new and economical methods of acquiring and transmitting data.

Data are needed to: detect up-to-the-minute traffic conditions in order to implement traffic control measures; supply driver and passenger information systems; and determine vehicle locations for fleet management. Different sensor technologies have been developed and are in use, *e.g.* loop and ultra-sonic detectors, but new technologies are required for more economical data collection, such as computer vision and intelligent vehicles.

For data transmission the most commonly used media are cables and radio waves. New developments in the field of communication systems are ISDN (Integrated Services Digital Network), digital radio with one-way or two-way communication, RDS (Radio Data System), cellular mobile radio, and beacon-supported microwave and infra-red communication. These systems provide new means of communication with drivers and passengers.

Summarising, the following conclusions can be drawn about the requirements for more efficient traffic management. Future action should include the management of demand for transport in the context of the overall transportation system – public transport as well as road and parking infrastructure.

The approach should be dynamic in order to provide traffic information to increase users' acceptance, and with it the effectiveness of the management system.

However, the system should remain easy for users to understand. Best results can be achieved if up-to-the-minute information is provided before the user starts his trip (*e.g.* at home), but it should also be provided during the journey.

In the light of this general strategy of traffic management, new concepts introduced in OECD countries have to be evaluated to see whether they can help achieve the objectives of demand management; their implications for cities and for society have to be addressed. We shall focus first on new developments in public transport, and second on those in the traffic management field. The emphasis will be on ICT (information and communication technologies) rather than on other technologies.

The structure of a computerised traffic-demand management system has been described by Mr. Hepworth (1990, see Figure 10) excluding fleet management and vehicle monitoring systems, in particular for public transport. Apart from fleet management, we can distinguish between information systems for drivers and passengers and control systems affecting traffic flow directly. Some of the measures introduced in OECD countries are described in the next section, and this is followed by an assessment of the perceived benefits and political implications.

Urban public passenger transport

An attractive – that means reliable, fast and inexpensive – service is a basic prerequisite for inducing the public to switch from private to public transport. It is becoming increasingly important to assess the reliability of transport systems, and a broad field is opening up for the application of ICT.

By far the most important application is in passenger information, which helps make public transport more accessible to people who are not familiar with it. In a few large and medium-size cities, information on the departure times of buses and trains is shown on electronic display boards. In some instances, delays and other disruptions in services are also displayed. This presupposes that operations are electronically monitored and that the display panels are connected to the control centres, or buses transmit data via radio.

Operation control systems have been under development for almost 20 years and are being increasingly used by public transport systems. They are meant to check adherence to schedules and to assist vehicle deployment, by warning of service disruption so that alternative transport can be provided as quickly as possible. These systems also provide passenger information as already mentioned. Another benefit is to help ensure good connections when passengers change from one vehicle or mode of transport to another.

Computer-assisted operation control systems (COCS) are used for bus, light rail rapid transit and underground transport. They are so widespread in OECD countries that it is pointless to single out any one system for particular attention. Innovative data transfer technologies now make it possible to send information between vehicles and their control centre safely

Figure 10. **Computerised traffic management systems with route guidance**

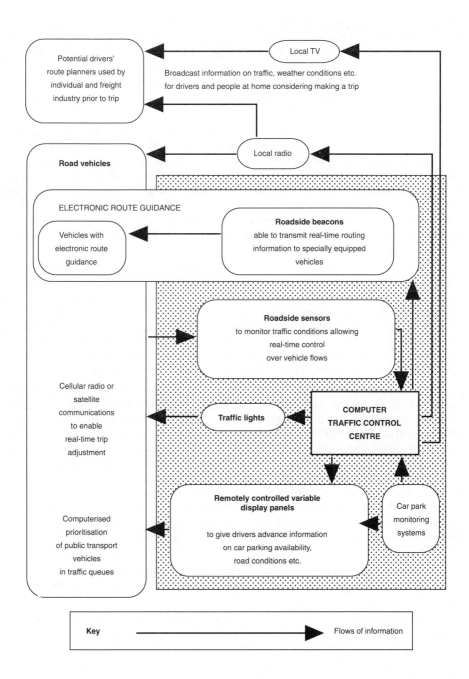

Source: Hepworth, 1990.

and quickly. COCS systems are indispensable for the customer-friendly operation of public transport.

Higher traffic density means that buses and trams are held up in traffic jams with increasing frequency. In many cases it is impossible, for financial and physical reasons, to run tramways in tunnels or on dedicated trackways, at least in inner city areas; alternative ways have to be found to run buses and trams more punctually. One way is to build bus lanes, but this is not possible everywhere; therefore systems for influencing the sequence of traffic lights are becoming more important – systems which would have been impossible without the use of modern communication technologies.

Typically, such a system uses infra-red to detect a bus when it is approximately 300 metres from the lights. It then checks what changes have to be made in the traffic signal programme in order to give the green light to the bus when it reaches the lights. Zurich has carried this development further than any other city, with the result that normally trams never have to stop at traffic lights. In this way, the average journey speed is considerably increased, which is to the advantage both of the passenger and of the transport operator, which needs fewer vehicles to convey the same number of passengers. The advantages of these devices are undisputed, and the restrictive effect on private traffic is minimal as long as the overall green time for all approaches to the signals is not changed.

The safest, most reliable transport service, and the one which saves most energy, is that which is operated automatically. It is not so much staff savings that are important, as the ability to use staff to improve passenger service. Transport systems with their own rail tracks are being designed to allow for automatic operation on later extensions. The most convincing proof of the soundness of this idea can be found in Lille, where the metro is operated totally automatically, and in Vancouver (Sky Train). A similar scheme is under way in Berlin, where an automatically controlled magnetic rail system is being tested on a 1.6 km section of line. These systems have demonstrated that passengers get used to automatic train operation quite easily.

Special demands are made on bus services in less densely populated (especially rural) areas. Owing to the low and often only sporadic demand for transport, it is normally not economically feasible to use large buses, as these tend to be driven around empty especially in off-peak periods. In the 1970s the idea was born of offering a passenger service based on actual demand. Various models have been developed for this purpose. In its simplest form, a service is run only when at least one passenger has contacted the operation control centre by telephone in good time. In a more sophisticated form, a computer assesses passenger requirements, but in this case too customers must have phoned in their departure and destination points in good time. If deployment is effected on-line, the passenger is told his departure time when he calls. Deployment is effected with a view to fast passenger service, but allows for general economic factors, i.e. routes are planned in such a way that as many passengers can be collected as possible.

Such a system is being operated in the South of France under the name of Christobald. There is a similar development in Greater Hanover, Germany, under the name Rufbus (Callbus).

A service provided by urban transport companies which is becoming increasingly widespread, at least in Germany, consists of asking the driver of a bus or tram to order a taxi to pick up a passenger at a designated stop. It is convenient for the passenger that a taxi be waiting for him when the bus or tram reaches the stop. This service, provided primarily for suburban areas, where and when no other public transport is available, is the result of close co-operation between the taxi industry and public transport operators. More and more people are taking advantage of this facility. In order to provide the service, the bus or tram driver needs a radio-telephone or data transfer connection to the operation control centre organising the taxi rendezvous.

One forward-looking development is in the area of passenger handling. After the replacement of ticket sales points by automatic ticket issuing machines as a rationalisation measure, passengers developed an antipathy to these devices, which are not very customer-friendly. It is now intended to remove this obstacle by introducing an intelligent ticket. A ticket the size of a credit card will be issued on the basis of the credit/debit procedure. It is proposed that in future the same card will be used to pay for other city services. Thanks to these smart cards, it will no longer be necessary to purchase a ticket or to look for the right change.

A similar development, applied to public transport vehicles, is the so-called "smart bus". In addition to a radio-telephone and a data transfer connection to the operation control centre, these buses are equipped with an on-board information system which automatically controls the announcement and display of stops in the vehicle. The ticket cancelling machines and/or reading devices for the smart cards are linked via this system. Passenger-friendly service is ensured by means of vehicle-related timetable supervision and the display of transport connections. The opportunities offered by the use of ICT are rounded off by the possibility of requesting green at traffic lights and, if necessary, of using a radio connection to display delays at the bus stops ahead, which therefore become "smart bus stops".

These examples show that there are many ways in which communications technology can be applied to improve the user-friendliness of public transport. Transport can be much more flexible because those who control it have much more, and above all up-to-the-minute, information available on current operations, and are able to convey information to the customer. Modern public transport is inconceivable without modern information and communication facilities.

Traffic control systems

Probably the most widespread form of traffic control is light signals. Although these are primarily intended to increase road safety at junctions, they have also considerably improved traffic flow. It would be beyond the scope of this summary to cover all the means of optimising traffic light control. In its most sophisticated form it is employed over large areas, in which lights are controlled by regional computers which are in turn monitored by a central computer. Thus all the prerequisites are fulfilled for changing from time-dependent traffic light operation to traffic-related operation. However, it is only possible to introduce traffic-related signal control if the necessary traffic data are available.

One key area of development of traffic light control was, and still is, co-ordination and optimisation of signals. This development has led to, among other things, the "green wave", which makes it possible to go through a series of traffic lights without stopping if a certain speed, which may also be shown on displays in the road, is not exceeded. Consistent use of this method has led to so-called city timetables, which ensure that the traffic flowing into the city in the morning and out of the city in the evening is handled as swiftly as possible. In some cases, so-called "doorkeeper" lights are activated, which deliberately hold up traffic at the outskirts of the city in order to prevent a traffic breakdown which would occur when the capacity of the inner-city road network was reached, and the harmful environmental effects that would result.

Today it is possible, using up-to-the-minute traffic data, to optimise traffic light control at individual junctions, allowing for the effects on neighbouring junctions, in such a way that the number of vehicles getting through the junction during one phase is increased. The possibility also exists of giving priority to public transport vehicles. A project with this objective has been successfully put into practice in a district of Turin. The time gains by trams and buses have led to a remarkable increase in journey speed. Time losses by private traffic have been reduced. Apart from the desirable effect on the traffic situation, this has also helped reduce environmental pollution and improve road safety. Results obtained to date show that the future of traffic light control without doubt lies in switching signals on the basis of actual demand. Using a similar approach, the GERTRUDE system was introduced in Bordeaux ten years ago and the experimental ZELT system in Toulouse.

Variable traffic and message signs play an important role in influencing traffic. If they are controlled on the basis of time intervals and/or the traffic situation, it is possible to display instructions tailored to the situation, such as prohibiting turning off in a certain direction, and to recommend alternative routes. Effective as these systems are, they have the disadvantage that they can only show a limited amount of information and, for reasons of road safety and because of city layout, only a limited number can be installed.

One specific but seldom used traffic management tool is reversible-lane signalling. This means that the lanes on multi-lane roads are allocated to different directional flows according to traffic volumes. For instance, on a five-lane arterial road, two lanes may be allocated to outbound traffic and three lanes to inbound in the morning, and vice versa in the afternoon. These systems make very effective use of road space when the traffic flow in one direction is significantly higher than in the other. These systems can, however, only be used to a very limited extent because of the complex design of the changeover areas and the number of signs required. In many cities in OECD countries they are used in tunnels, often to divert traffic when certain lanes are partially or totally blocked.

Because of the increasing density of traffic and the greater probability of congestion, congestion warning systems are being introduced on busy roads, especially urban motorways. Their main aim is to increase road safety, but they also serve to maintain and speed up traffic flow, as long as traffic does not come to a standstill. As soon as the local data recording devices register traffic disruption, the speed limit is reduced and the display boards are switched to show congestion. By reducing the speed limit, traffic throughput can be increased to its maximum. However, if traffic demand is beyond road capacity, these systems have proved highly effective in making alternative route recommendations. Future developments must pay more attention to weather-related road conditions so that car drivers can be more effectively warned about fog and ice.

The ability of modern communication technologies to exchange data with moving vehicles provides a good basis for setting up toll stations. In contrast to the Italian and French toll motorways, most roads do not have sufficient space to install toll stations. However, the latest transmission technologies enable road-use charges to be calculated for moving vehicles. This opens up a wide range of ways of collecting charges, at least from a technological point of view. This essay will not go into detail about the social consequences of introducing road-use charges (road-pricing), but it should be pointed out that more and more people are expressing the opinion that, in view of the increasing traffic and the limited capacity of our cities to accommodate motor vehicles, restricting traffic by means of charges should be considered at least for limited times and areas. This method of determining charges is planned for Oslo and Trondheim in Norway. Further applications are currently being planned, for example in the

Netherlands. Interest in the use of systems centres mainly on motorways and expensive structures, *e.g.* tunnels and bridges.

The contribution that automatic charging systems can make to improving traffic management by helping to reduce traffic and equalise it between peak and off-peak hours is considerable. This is primarily because most car drivers are only prepared to alter their driving habits when not doing so involves considerable costs. A highly desirable side-effect from the point of view of traffic policy is that the revenue from charging can be used to improve public transport.

Another practice which has proved its worth is automatic traffic monitoring. This includes monitoring adherence to parking time restrictions. In Germany, for example, a mobile data-collection unit has been developed, by means of which infringements can be entered manually into a hand computer, from which they are automatically processed up to the point of issuing a warning and fine. Future systems for monitoring parking times are likely automatically to register arrival and departure times, and to book charges, *e.g.* via a smart card.

Driver information systems

Regular traffic warnings via radio stations have not yet been introduced in all OECD countries. Radio warnings have their limitations, which are reached when the number of reports issued (*e.g.* at weekends and in holiday periods) becomes too great. For this reason, developments for the future will be aimed at providing more regional information and greater differentiation in traffic reports. This presupposes, however, that comprehensive, up-to-the-minute, reliable information is available on the extent and duration of traffic disruption. In future it will be possible, via the Traffic Message Channel (TMC) of the Radio Data System (RDS), to transmit traffic reports to vehicles on a regional basis. This will only work, of course, if the car radio is equipped to receive the channel and if radio stations are equipped with the necessary technology to transmit reports via the channel. In future the car driver will therefore receive less irrelevant information, and at the same time the reports he does receive will be more comprehensive and up-to-date. The advantage for radio stations will lie in the fact that their programmes will no longer need to be interrupted for traffic reports, as these will be transmitted in coded digital form.

Traffic information is also displayed using roadside equipment in many OECD countries. The amount of information displayed is even more limited than that which can be conveyed by radio but it is provided to all road users.

In a new development, traffic information can be transmitted directly to individual vehicles in either visual or spoken form. To achieve changes in motoring behaviour, such systems have to transmit information relating to the current traffic situation. However, this is only possible if there is communication between vehicles and the roadside.

A field test is currently being carried out in Berlin with this aim in mind. Some 650 vehicles have been furnished with on-board equipment, by means of which route recommendations based on the traffic situation can be received and displayed in the vehicle (Figure 11).

The recommended route to the destination, keyed in by the driver at the beginning of the trip, is indicated by arrows showing the driver when and where to turn left and right. The vehicle is capable of determining its position using information from the odometer and a compass. Based on this information, the vehicle is then navigated to its destination via the fastest route. Communication with a central control computer is effected by means of so-called "beacons", which transfer the data necessary for navigation to vehicles without identification

Figure 11. **LISB route guidance system**

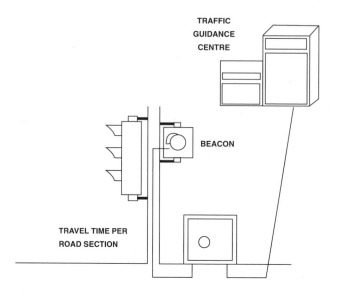

TRAFFIC
GUIDANCE
CENTRE

BEACON

TRAVEL TIME PER
ROAD SECTION

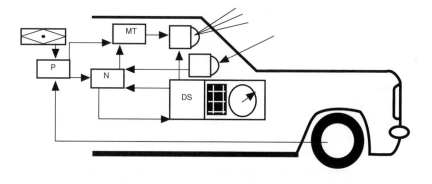

P   Position finding device
N   Navigation device
DS Destination store
MT Travel time measuring device

in monologue infra-red form. The special feature of this system is that the vehicles themselves assist in the recognition of traffic disruption by registering their travel times at traffic lights, and when passing a "beacon" by conveying this information anonymously to the control centre. This data provides the basis not only for optimised route planning, but also for a number of other applications such as optimising traffic light control according to recorded congestion.

The Berlin Route Guidance and Information System (LISB) points to a development which can be expected in Europe and Japan, and perhaps in the United States and Canada too, in the second half of the 1990s. It can be seen in the AUTOGUIDE system in London and the SOCRATES system being developed within the European DRIVE research programme.

Dynamic route guidance and traffic information systems like LISB and AUTOGUIDE will be used for traffic management in future. However, it is important now to develop the strategies necessary to use these systems to improve traffic handling in our cities. By planning routes centrally, it is possible to direct traffic past congestion-risk areas and so to make more intelligent use of the road infrastructure. Precisely because these systems are of so much benefit to municipalities, every effort must be made to ensure that they are operated by public bodies and actually implemented in order to improve traffic management.

Developments similar to LISB and AUTOGUIDE are being pursued in Japan. In Tokyo, two systems named RACS and AMTICS are being tested, which act as navigational aids to the driver and are intended in future to warn of traffic disruption. In contrast to the LISB system, which navigates the driver to his destination by means of simple signs on a display, the Japanese systems show sections of networks on a monitor and mark the disrupted sections in different colours, depending on how serious the disruption is. A communication link is provided using beacons or cellular radio. As no route recommendations are given with this system, the effect in terms of influencing traffic is not as great as with the LISB and AUTOGUIDE systems. Nevertheless, this development shows the importance of control and communication systems.

Using the data provided by LISB and AUTOGUIDE, an individual traffic information system is being developed in France which shows the degree of traffic disruption on any route selected by the driver on a display in the vehicle. In this way the driver has the opportunity to choose an alternative route if necessary and will receive updated information via RDS while driving. This system, developed under the name INF-FLUX, is currently being prepared for introduction in Paris.

The effect of control systems is considerably increased if they are linked to parking guidance systems. In this way, the amount of traffic looking for parking space is reduced. These systems are particularly effective when route recommendations only include those car parks which still have free spaces available. In the medium term, it may become possible to reserve a parking space from a moving vehicle. In this case, accounting would have to be effected in cashless form – yet another possible use for the smart card.

Many OECD cities have successfully introduced parking guidance systems that inform the driver about available parking facilities. Cologne uses a system by means of which the car driver is also shown the number of parking spaces still available. This makes it easier for him to decide on a particular location. However – and this is the drawback of most systems – they are of no assistance to the car driver when all parking facilities are occupied. It is therefore always advisable to direct traffic flowing into a city to car parks at the city outskirts when parking facilities are exhausted, and to convey travellers comfortably and quickly from there to the city centre by frequent public transport services. In future, park-and-ride systems enabling drivers to park easily and continue their journeys by public transport will have to be considered more intensively as part of traffic management systems.

Fleet management systems

The increasing costs of running vehicle fleets create a need for more effective planning and flexible operation. Traffic congestion and a move towards precisely timed deliveries make this even more necessary.

Fleet management demands that the locations of the vehicles are known to the dispatch centre. This allows the operator to allocate additional orders to the best-located vehicle via radio, minimising cost. Even more efficient operation can be achieved by making use of travel time information acquired from experience for optimal route planning, taking into account variations in traffic volume. Special attention must be given to the transport of hazardous goods; routes have to be selected which enable vehicles to be monitored and rescue plans activated.

Taxi operation applies the principles of fleet management to passenger transport. In its most advanced form it allows calls for cabs to be passed to the vehicles automatically. When a customer calls in, his order is keyed into a computer which keeps track of the position of all available taxis. The computer selects the vehicle closest to the pick-up point and supplies all necessary information to the driver. Such systems are in operation in many OECD countries such as Sweden, Japan, the United States and Switzerland, and under development in others.

Although primarily designed to optimise fleet operation, fleet management systems also help to use road capacity more efficiently. Therefore these systems have to be included in a traffic management scenario.

Having outlined the various means of achieving better traffic management – most of which are based on ICT – this essay will now focus on the impacts and implications in some detail.

Assessment

The examples described so far concern developments in public and private transport which can be seen in operation. The Zurich system for giving buses and trams priority at traffic lights has become a model for many European cities. It was developed by the city and is operated by the city police with great success. Although it was not designed for motorists, it does have some impact on them. The system has been accepted by the inhabitants of Zurich. The adaptation of such traffic regulation measures depends on the public's attitude. Hence each city must define its own strategy for devising, implementing and operating such systems. The same is true of automated rail transport.

In-vehicle information and communication systems offer a variety of ways of improving traffic efficiency, with benefits for the individual driver and for the community. There is a big move worldwide to go for systems developed by car manufacturers and the electronics industry, recognising that the future will bring more intelligent cars. Therefore cities have to be prepared to take up this challenge and to integrate the smart-car development into an overall urban traffic management scheme.

As traffic outstrips road capacity, particularly in city centres, restrictions are being discussed more and more. The cost of congestion is being paid by the community, not by those who cause it. Technically it is feasible to charge drivers for using roads, with the price depending on location and time of day, as has been demonstrated in Hong Kong. However, social considerations (fairness) and limited spare public transport capacity were the main reasons that road pricing was not introduced there.

As mentioned at the beginning of this paper, because private and public transport share the roads, it will become increasingly important to influence traffic flow and behaviour in an integrated way. This presupposes, however, that traffic-related and operational information is collected at a central point. With this aim in mind, consideration is now being given to the benefits which would be achieved from a control centre covering all traffic on the roads operated privately or by public bodies. One advantage would be that a control centre could initiate tactical measures to influence traffic behaviour, in advance of a problem occurring, and thus become a more proactive than reactive instrument of traffic management. However, comprehensive studies are still required to assess the possibilities and to determine the responsibilities of such a control centre.

Although the idea of park-and-ride is not new, it is gaining in importance as a result of increasing differentiation between the roles of private and public transport. The masses of vehicles which will be converging on our cities in even greater numbers in the future can only be satisfactorily handled if there are sufficient transfer facilities from car to public transport at suitable points. In this, too, information technology will play an important part, as a car driver will not be prepared to leave his vehicle and continue his journey unless he has been well informed about available public transport.

Great faith is being placed in traffic information provided before the start of a journey, as a means of influencing traffic behaviour. Systems such as MINITEL in France should in future be increasingly used to disseminate traffic information and to recommend means of transport. In many cases the car is preferred to public transport solely because road users do not know how to get to their destinations by public transport, which service they have to use and how long the trip will take. Making this information clear is certain to lead to greater acceptance of public transport.

Too little attention has been given in the past to the inclusion of goods and commercial traffic in traffic management. New forms of deployment planning and new logistical concepts are needed to gear goods traffic better to urban needs. Taking into account fluctuations in travel time at different times of day, delivery journeys can be planned in a much more cost-effective way than is generally the case now.

RECOMMENDATIONS

As in other areas of public life, the rapid development of ICT has helped devise new ways of influencing traffic. Because of the growing traffic problems in our cities, it is essential to make use of these technologies to improve road travel and the efficiency of transport companies.

In future, public transport companies will not be able to operate without computer-assisted control systems. This will also apply to improving passenger information about the current operation of services and the timetable, including alternative travel possibilities and fares.

Figure 12. **Traffic control measures**

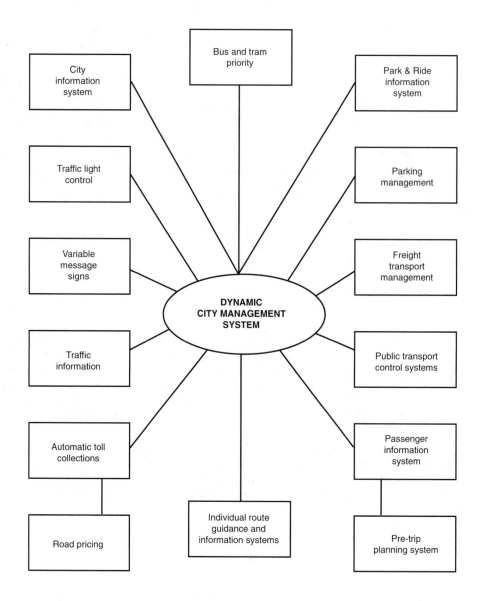

Transport systems will also need to be able to influence traffic lights if public transport is to be given priority over private traffic, as is politically seen to be necessary.

On top of this, more work is needed on influencing private traffic. Apart from continuing the development of collective traffic influencing systems, innovative leaps forward are to be expected in terms of influencing drivers individually. In the second half of the 1990s, the intelligence in the vehicles themselves will have advanced so far that individual traffic guidance and information systems will be offered by car manufactures. Here lies a great opportunity, as these systems will make it possible to influence traffic in a much more differentiated way than can be done collectively. In order to supervise these developments, and to use the potential of the systems to achieve political and environmental aims, it will be of paramount importance for cities and municipalities to allow access to the traffic control centres.

As has been shown in all the examples mentioned, information and communication technologies have expanded the means of controlling and diverting traffic as well as affecting user behaviour and transport mode choice. Therefore it will be possible in future to apply control measures individually tailored to the special needs of cities (Figure 12).

Traffic management should be seen as an integrated process, including managing demand in order to influence journey planning and mode choice, the better to use the overall transport infrastructure. In this context, co-operation is to be desired between transport providers, rather than competition and confrontation. Cities must devise traffic management strategies setting out policy options that will help protect the environment from the undesirable impact of traffic.

BIBLIOGRAPHY

ASSOCIATION OF ELECTRIC TECHNOLOGY FOR AUTOMOBILE TRAFFIC AND DRIVING (JKS) (1989), *International Survey of Automobile Information and Communication Systems,* Tokyo.

COMMISSION OF THE EUROPEAN COMMUNITIES, (1985), *Electronic Traffic Aids on Major Roads,* Directorate of General Science, Research and Development, Brussels/Luxembourg.

DEPARTMENT OF TRANSPORT (1986), *Autoguide: A Better Way to Go?* London.

ECMT, Ministerial session (1990), *Transport Policy and the Environment,* Paris.

HEPWORTH, M.E. (1990), *Applications of Information Technology in Transport,* University of Newcastle upon Tyne.

OECD (1987), *Dynamic Traffic Management in Urban and Suburban Road Systems,* Road Transport Research, Paris.

OECD (1988), *Cities and Transport,* Paris.

SPARMANN, J.M. (1989), *LISB Route Guidance and Information System: First Results of the Field Trial,* VNIS Conference, Toronto.

SPARMANN, J.M. (1990), *Urban Traffic Control: Contribution of Route Guidance and Information Systems to the Solution of Traffic Problems,* Strassenverkehrstechnik 34.

USE OF INFORMATION AND COMMUNICATION TECHNOLOGIES IN THE FIELD OF ENVIRONMENT AND ENERGY CONSERVATION

Lorenzo CASSITTO
Professor at Politecnico di Milano University
Sergio SEGRE
Consultant
ITALY

Lorenzo Cassitto graduated in mechanical engineering from the Politecnico di Milano University. He is currently associate professor in the Department of Energetics at the Politecnico di Milano, where he teaches courses in applied physics and in heating and air conditioning. He is also president of his own consultant company (Artea Sarl, Milano).

Prof. Cassitto is the author of over 80 scientific papers, and over 50 other publications. He is also referee at the Committee for the European Development of Science and Technology, European Economic Community, and Italian Delegate to several Commissions of the EEC, the OECD and NATO/CEMS.

Sergio Segre has a degree in law from Rome State University and Vatican Lateranense University. For many years, he was responsible for public technical communication for the Olivetti group. As president of CIDE Consulting (Corporate Image Development and Strategic Consultant) and president of SCS (Communication Strategies), he was responsible for the computerised environmental simulation projects developed at the request of the Italian Ministry of Environment.

He is also an advisor for international public communications to Regione Toscana for Proggetto Etruschi, to the environmental utilities of Ravenna and Modena, and to Milano's Municipal Energy Authority, for institutional communications and image.

STATE OF THE ART: FIELDS OF APPLICATION

Introduction

The state of the art in any form of human activity is never precise or totally up to date because, at the very moment in which it is being described, something new is taking place which makes the account out of date.

This is particularly true in the case of information and communication technology, and the changes that occur from one day to the next are sometimes extremely far-reaching.

The application of this technology to the fields of energy saving and environmental protection has been going on for years. Some solutions have been tested and proved while other experimental applications have not, at the time of writing, reached a conclusion.

There are no absolute values or reference points for energy saving and environmental protection, because the former depends on what standard of living is demanded and possible – and this varies from one country to the next – and the latter depends on how successful we are, technologically and economically, in foreseeing and controlling the consequences of using the energy needed to produce the standard of living demanded.

Once objectives have been established, it is essential, in order to achieve them, to have the understanding and co-operation of the public. Information and communication technology is the most effective instrument to explain the complexity of energy and environmental problems to the public, and to encourage their involvement.

It would in fact be absurd to expect people to get involved if they had not been provided with all the information to enable them to understand the purpose and potential results of their efforts and, if necessary, their sacrifices.

Information technology makes it possible today to manage extremely complex systems – which is what modern cities are – with which conventional technology would be unable to cope. Cities today are in a predicament, faced on one hand with a growing need for energy and services, which produce new pollution, and on the other hand the absolute need to protect an environment that is increasingly threatened. Information technology and the new techniques that it makes possible allow the city system and its way of life to be reorganised, and to move towards achieving expectations of a higher standard of living.

Great care must be taken when deciding objectives, which must be consistent and quantified; objectives that are excessive rapidly become unattainable, and turn all endeavour for their achievement into useless effort.

Applications

Modern cities – and the bigger they are the truer this is – constitute a complex system of correlated functions and spaces that impinge on one another: transport, heating and air-conditioning, waste disposal, industrial and commercial activities and related services.

The majority of these activities consume energy. Vast quantities of primary energy resources – oil, natural gas, coal, uranium – are burned, with grave harm to the environment that can only be reduced by the most up-to-date clean-up technology.

In these circumstances, the personal opinion of the authors is that environmental protection and energy saving can only be brought about by the integrated management of the city system, in the hands of a small number of city authorities which have the necessary powers to lay down effective programmes of action and see that they are carried out. The main sectors covered by these public authorities are energy, transport and environmental services, to which must be added town planning, which traditionally comes directly under the municipality.

These departments must work in close contact to ensure the combined effectiveness of their programmes, rather than allow them to conflict with one another. Take the case of traffic and urban heating, the two main causes of pollution of the air that we breathe. Only by having joint, co-ordinated programmes can the city environment be reclaimed.

Information technology makes it possible to control individual sectors and the contacts between the various departments, using programmes and procedures that are already defined or partly in operation in certain cities, and still being studied or at the experimental application stage in many others.

New monitoring and control techniques, and the study of new highly efficient systems for supplying energy in cities, which got off to a promising start immediately after the first energy crisis in 1973, have slowed down in recent years, because of the low price of oil. Even research into the new, intrinsically safe nuclear reactors has been reduced as a result of the availability of cheap oil.

The emergency facing us today is dire, both because of uncertainty surrounding oil supplies, which remain the main source of energy for many countries, and because of the threat to the environment, which in some cities has exceeded tolerable limits.

When planning new solutions for cities today, the protection of the environment is the main objective, more so even than energy saving. The quality of the air, water and soil is no longer only a constraint that must be observed, it is the prime consideration. Principal examples of the application of new information technology combined with technological innovation are as follows:

District heating and energy recovery

District heating consists of distributing to the entire city or part of it heat produced in a single plant or a small number of plants.

The plants can be equipped either with multi-fuel boilers, capable of burning various fuels, or with systems for recovering waste heat or exploiting low temperature sources (heat pumps).

Automation and control systems allow the available energy – from primary or secondary sources – to be used in the most effective way.

Heat production is based[1]:

- firstly, on the recovery of waste or free heat from industrial plants or waste incinerators[2];
- then, from the most economic sources (coal, peat, with emission purification), subsequently from highly efficient plants (heat pumps, combined cycles); and
- finally, to cope with peaks of demand, from the highest-valued fossil sources (natural gas, oil).

Heat produced is used at a rate related to its temperature. This is possible thanks to "intelligent systems" which can manage the production of heat in relation to users' needs, environmental conditions and the state of the equipment (power stations, distribution networks, sub-stations), processing the data automatically in real time, in integrated mode. Information technology makes possible the accurate and at the same time easy management of the thousands of consumers connected to the network (monitoring, metering consumption, invoicing, making connections to the mains, carrying out maintenance *etc.*) but it does not constitute per se a means of improving the efficiency of the plants. In this case it functions as a system of regulation and control.

District heating[3,4,5,6,7] is very widespread in northern Europe, particularly in Sweden, Norway, Denmark and Finland (some cities have 90 per cent of their heating provided by this method) and is general in central Europe – the Netherlands, Germany, Austria, Switzerland, France, Hungary, Czechoslovakia and the Soviet Union. It is beginning to be introduced south of the Alps, for instance in some cities of northern Italy, such as Brescia, Verona, Milano, Torino and Reggio Emilia. The complexity of the heat production plants differs according to the cities that they serve.

One particularly important example is the new urban refuse incinerator in Goteborg, Sweden,[8] which not only solves the serious urban health problem that refuse presents but also makes significant energy recovery possible (district heating and co-generation) by using highly efficient new technology (absorption heat pumps that harness the heat of the water used for washing the flue gas).

New energy systems: high performance heat pumps

In Japan, as part of the energy-saving programmes of the Ministry of Industry and Trade (MITI), a "Super Heat Pump Energy Accumulation System" project was started in 1985 and developed by the New Energy and Industrial Technology Development Organisation (NEDO)[9]. The budget for this is $80 million.

Its objective is a new system, consisting of a compression heat pump with heat accumulation, for the air-conditioning of buildings, and for district heating and industrial processes, which is able almost to double the output of energy compared with equivalent machines now on the market. For the heat pump alone, the aim is to raise the co-efficient of performance (COP) to values of 6 in the case of water production for district heating, 8 for hot water with heat accumulation and 3 for high temperature heating. In this last case water is used as the working fluid, with a turbo-charged compressor. During the research, importance has also been placed on the adoption of new working fluids that are not dangerous to the environment, or fluids with special physical characteristics (non-azeotropic mixtures).

The experimental stage, relating to the study of the individual components and the opening of a pilot plant with a thermal output of 100KW, has already been completed. A 1 000KW system is now being tested, and it is planned to implement a 30MW plant by 1992. A plant of this size is commensurate with the needs of the average Japanese office building, which has a surface area of about 150 000m^2.

"Intelligent" building

This term is used to describe buildings which have a computerised system to regulate, supervise and collect data on all their technical equipment as well as safety factors and other services with which we are not concerned here.

The system can be installed locally or at a remote site connected to the building via a communication network. The typical architecture of a remote processing system of this kind is as follows:

- a local control and monitoring system in each building;
- a centralised regulatory and supervisory system installed at a service centre, equipped with appropriate software.

The purpose of the energy management system is to obtain the predetermined level of comfort, whilst minimising consumption, and it provides for the following:

- optimised start-up/shut-down, in relation to previous operating data;
- control of total heat consumption;
- night operation;
- cyclical shut-down;
- limitation of electricity consumption peaks.

The remote management system of the Azienda Municipalizzata Case Popolari e Servizi (AMCPS) of the Municipality of Vicenza, Italy,[10, 11] is described in Annex 4.

These systems are widespread in OECD countries, and in France in particular, where they are used not only for controlling the energy functions in buildings – this is particularly common – but also for all the other functions.

Telemanagement systems of urban services

The aim of the telemanagement systems is the monitoring, controlling and operating, based on computer programmes, of one or more urban services, such as for instance the supply of water, electricity, natural gas *etc.* to improve performance and reliability.

The following two examples explain how it is possible to make use of such systems.

Milan

The Azienda energetica municipale (AEM) of Milan, Italy, the city-owned energy authority, has developed a telemanagement system for the natural gas supply to the city.

The distribution network is based on high-pressure mains which feed a number of pressure reduction stations. From each of these a piped local distribution system serves consumers.

The basic concept of this system is that:

- every plant is equipped with a tele-operated control device that performs a check cycle 12 times a minute;
- the data are transmitted to the AEM remote central control room via a communication network (dedicated telephone lines or fibre optics).

This telemanagement system makes it feasible to rationalise the natural gas supply, to increase its safety and reliability, to reduce the number of employees and the operating costs, and finally to save energy. The cost of the system from 1976 to the present day has been approximately 2 billion lire. When the project is complete the telemanagement system will control 160 pressure reduction stations, 70 per cent of which are already fully equipped and connected to the system.

The AEM is also developing a computerised system and procedural scheme able to manage the calls of the users, rationalise the work of the maintenance teams, and create a historic database with the object of identifying pipes, pumping stations and buildings which have caused problems (contact: Dr. Bartolini, AEM, telephone 0039/2/77201).

Istanbul

The Water Supply and Sewerage General Directorate (ISKI) of Istanbul (Turkey) uses a telemanagement system to improve the quality of its service. ISKI uses the Supervisor Control and Data Acquisition System (SCADA), through which it is possible to monitor the water supply.

The concept is very similar to that of one of the AEM systems; so far 57 Remote Terminal Units (RTU) have been connected to SCADA check dams, pumping stations and the distribution network. Operating and checking data are transferred via a transmission network.

The telemanagement system described serves the following purposes:

– to control the amount of water distributed to different sections of Istanbul;
– to take action in cases of fire, flood, *etc.*;
– to save energy.

The approximate initial investment was $5.5 million (contact: Kayhan Kelginlioglu, telephone 0090/11/1325085).

Air quality monitoring networks

Monitoring air quality is one of the most significant applications of the new information and communication technology. The example of Milan Province, which is the largest in Italy, is described below[12], [13].

This network for capturing air pollution data has been in existence since 1958 and is managed by the "Presidio Multizonale di Igiene e Prevenzione" (PMIP). In 1973, the network was equipped with remote measuring systems which were used for automatic data management by means of a computer connection direct to the on-site instrument centres. The network consists of 41 stations for measuring the main pollutants such as SPM (solid particulate matter), SO_x, NO_x, and O_3 and 12 stations that also measure the meteorological parameters (wind speed and direction, air temperature, pressure, humidity). Parts of the instrument centres are managed by other organisations (AEM, ENEL, BASF, IP *etc.*) but they are all interfaced with the PMIP computer.

The individual instrument centres are equipped with a local minicomputer which improves the reliability of the system, whilst the central computer manages data transmission, checks the recordings taken and processes the measurements in order to make them comparable with the existing standards on air quality, and finally serves as a databank.

The efficient functioning of this network makes it possible to show in real time the extent of the atmospheric pollution over Milan, which represents one of the most critical environmental problems of that city, and to take whatever immediate steps a particular situation requires. Depending on the concentration of pollutants, the municipality operates measures ranging from public warnings to reduce the power consumption of space heating plants to, as a final resort, prohibiting the use of private cars. A future development will be the use of dedicated telephone lines to communicate between the main centre and the measuring stations, and in addition video monitors are being installed to provide the public, in real time, with information on the quality of the air.

Some instrument centres for recording noise are also to be installed. In terms of its structure and size, the Milan Province network compares favourably with the one in San Francisco, which is one of the most important in the world.

Information systems on clean technologies

Denmark is close to creating an information system to improve environmental pollution control.

In accordance with the Ministry for the Environment's "Clean Technologies Programme" the Clean Technologies Centre of the Copenhagen Technological Institute is developing, with the aid of some Danish firms, 200 index cards on the iron and steel industries. These cards collate all the main data about the industries' clean-up plants and describe the processes themselves. Similar projects will be developed for the food, woodproducts and furniture manufacturing industries.

The project also includes setting up a centralised online database; users will be able to consult it and then print the information required. The National Agency for Environment Protection will set up a special office to manage this very complex system.

The database is on two levels. The first provides brief information on:
- technical data about clean technology advantages;
- economic data;
- environmental data;
- references and bibliography.

The second provides detailed reports on the technical, economic and environmental features of the clean technologies, compared with current technologies, for a specific manufacturing process or industrial sector.

The cost of the system development has been estimated at $1 million; a similar amount will be necessary to extend it to the other main industrial sectors. The remaining costs (office, management, maintenance) are estimated to be $500 000 when the system is fully operating.

Computerised traffic management

The fundamental aims of traffic management are to improve the quality of public service, to reduce energy consumption and, most important, to protect the environment from noise and pollution.

Computerised traffic management can be put into effect either by controlling traffic lights or by intelligent monitoring of the public transport system. Operationally, this means recording the number of vehicles per hour passing a specific point in the city and their speed, using photo cells and radar monitoring or, in an emergency, police radio. The recorded data is transmitted to the central traffic management system, and the phasing of the traffic lights is set accordingly.

The system achieves its objectives for the most part by giving priority to public transport, through special reserved lanes, and through traffic light control at crossroads.

In order to achieve the primary aim of efficiency in public transport, which is the best way to convince private individuals to stop using their cars, the service must be punctual and comfortable.

In some cities additional measures have been adopted. Examples are: the rationalisation and relocation of points of sale (shops, supermarkets *etc.*) in order to provide a better distribution, in time and space, of the flow of trade vehicles and private cars; the electronic payment of tolls by equipping vehicles with transmitters which send signals that are picked up as the vehicle passes through the toll point and passed to the central control system which debits the amount of the toll, without the vehicle having to stop in order to pay the toll; and, finally, an electronic system for selecting the best route to a pre-determined destination.

Lastly, in order to avoid congestion, optimise consumption and reduce atmospheric pollution, computerised systems have been developed that ascertain and display, on luminous panels placed at city entrances, details of car parks with spaces available (Cologne, Germany, and Bayonne, France).

To assess the importance of traffic regulation, reference may be made to the most recent data from the European Community[14].

Every year 50 000 people die in road accidents, 1.7 million are injured and 1.5 million are left permanently handicapped. The cost of traffic congestion is 500 billion ECUs, and the cost of pollutant emissions from vehicles, which contribute significantly to the total pollution, is placed at between 5 and 10 billion ECUs. The Committee of the European Community's "Drive" programme has calculated that if the electronic route selection system were installed on 25 per cent of London's vehicles there would be a saving (in terms of fewer miles travelled and fewer delays) of about 100 million ECUs per annum. More or less complete traffic control systems have been adopted in many countries belonging to the OECD.

The Los Angeles programme[15, 16]

Los Angeles in California has a population of 12 million; atmospheric pollution is severe, especially dust, sulphur and nitrogen oxides, photo-chemical smog and lead. The situation is aggravated by the morphology of the territory. A first action plan, in 1982, based on the use of alternative energy sources, advanced technology and control of the management of the territory and transport, was not adopted because it did not guarantee that the specific standards would be achieved.

The subsequent "Air Quality Management Plan" (AQMP) issued in 1989 is designed to operate in three stages:

– First, improvement of air quality monitoring technology and the installation of new monitoring and control systems (particularly in relation to everything that generates ozone).

- Then, greater use of the available technology and vigorous application of the traffic regulations.
- Finally, the development of new technologies, with the aim of eliminating combustion as the power source in vehicular transport. In practice, the plan proposes converting 40 per cent of cars and 70 per cent of trucks and buses to methanol by 1998. Subsequently, by 2007, it is proposed to withdraw petrol and diesel vehicles, and have 95 per cent of the cars on the roads electrically powered.

It is obvious that the new information and communication technology plays a fundamentally important role, because it would not be possible to pursue the above objectives without having recourse to systems capable of managing enormous quantities of data and sophisticated mathematical models.

Information

The very size of the problem of the environment, energy and big cities, and the vast numbers of people involved, have made public participation an essential element in the implementation of any programme of environmental protection or the safeguarding of the quality of life in large conurbations.

Environmental problems are however very complex, and their equilibrium depends on an increasingly large number of variables. It is difficult to identify by instinct alone what should be the correct way of life. How should we set about this?

In the past, the rules of social conduct, like the rules of personal behaviour (good manners), were simple and easy to learn. Today good intentions and goodwill are not enough. It is necessary to understand the problems in all their complexity, because this is the only way for us to determine how we should behave, and how to co-operate to defend the quality of life, or to insist that the authorities take adequate measures, which we then monitor.

It is therefore necessary to use promotional and educational techniques to improve public understanding, and this is only possible if new communication techniques and new instruments are used.

We have to go back to the origins of science: direct experience is the mother of knowledge, the guide to behaviour.

In a situation where environmental equilibrium, especially in towns and cities, is becoming increasingly fragile, it is too complicated to learn by heart all the ways in which we should behave and react in every possible situation; equally, it is impossible to analyse individually every action in our daily lives, to make those actions more consistent with the ever-increasing restrictions that are imposed on us today. Our behaviour must be automatic, a conditioned reflex.

But this in only possible if we can come to understand the logical or random outcome – the consequences, that is – of whatever behaviour we adopt, through direct experiment such as laboratory research. Information technology today offers us the possibility of this new, effective method of learning, with immediate, direct knowledge obtained through simulation programmes.

If these programmes extend beyond the purely scientific field and become "games" they will lose none of their validity, and will succeed in interesting a vast public and in involving people in the difficult situations that have to be tackled. They will be useful, therefore, to the public, politicians and public authorities.

A study carried out by Columbia University on energy strategy for the 1980s and 1990s, commissioned by Con Edison of New York, recommended amongst other things that investment in public information services should be increased, since information and knowledge of the problems were increasingly necessary for the involvement and participation of the public.

This paper will look in detail at some of the results of interactive computerised programmes implemented in Italy by public energy authorities (Milan) and the Ministry for the Environment, in relation to the following:

- air pollution in cities (AEM Milan);
- reclamation schemes for polluted rivers (River Lambro – Ministry of the Environment);
- the fight for the protection of the environment (Gulf of Naples – Ministry of the Environment).

It will also examine experiments in methodological effectiveness using instruments and advanced information systems developed in the United States, in regard to energy conservation and environmental protection in New York City and Los Angeles, and in other cities.

Institutional analysis

The fundamental technological changes are characterised by objectives, implementation times and a rate of obsolescence that are without precedent. However, cities must commit themselves to policies that involve using considerable resources for long periods of time, and this involvement is irreversible.

The magnitude and uncertainty of these developments constitute a new challenge to city programmers. There are two key questions:

- how to programme irreversible commitments of large resources, in a continuously developing situation that is consequently highly uncertain;
- how to exploit the most recent technological developments in order to give better service to city residents and cope more efficiently with the needs of a constantly changing society.

The first question concerns investments in the infrastructure which cannot usually be carried out in isolation, and are very costly and interdependent. Their physical lifetime tends to be long.

The volatility and scale of recent demographic and technological change mean that decisions to invest in infrastructures have to be taken in a climate of constantly increasing uncertainty as to cost and the possibility of error.

The second question concerns the rational use of the potential offered by new technology. This technology offers opportunities to speed up economic development, to facilitate economic transactions, to increase the efficiency of energy usage and to promote energy savings in order to improve environmental and working conditions, and to define new models of social interaction.

At the same time it enables urban administrations to develop new, more flexible services, and to become efficient. But, in order to provide multifunctional technological and information systems, cities must still devote very large resources and investment to infrastructures such as wiring and the construction of interactive networks inside and between buildings, and throughout the city. However, the premature allocation of large resources can lead to costly errors if the forecasts on which they are based do not materialise, or if the selected technology immediately becomes obsolete.

There is no general solution to this dilemma. The right answers depend largely on local factors. In general terms, the method of planning should be something on the lines of: "What changes should the city provide for in advance by using resources in a long-term perspective, and what changes should it tackle on a contingency basis?".

Wide areas of urban policy are involved – land use and the planning of public space, the multifunctional capacity of existing structures, basic infrastructures, economic development, services, environmental protection, local energy supply systems, appropriate residential structures, the definition of zones and regulations, communication networks, traffic models and models of public transport systems.

This is why town planners and policy makers must be backed up by research into the spatial nature of technological and socio-demographic changes and their implications for planning.

It should be pointed out that national and local authorities have undertaken fairly comprehensive initiatives in relation to protection of the environment and energy saving, with both short-term measures designed to resolve immediate problems (reduction of pollutants) and long-term measures designed to resolve more general problems (reduction of the sources of pollutants).

The following examples describe the action undertaken by countries of varying resources, culture and geographical collocation to manage environmental protection and conservation:

Italy

On 3 August, 1990, the Italian Parliament approved a three-year programme of environmental protection throughout the country. It covers the following: wastes, water, air and noise, scientific research, protection of wildlife, industrial risks, production of a national geological map, information and education.

The programme outlines the strategies to be followed to solve the main problems and specifies the funding required: 45 billion Italian lire for monitoring and information systems in areas of high environmental risk, 10 billion lire for the development of information systems using new technologies, and 2 billion lire for the creation of a centralised data bank.

Coupled with effective environmental protection, this programme is designed to increase public sensitivity to the problems of the environment. It is also intended to co-ordinate the efforts of national, regional and local bodies.

Norway and New Zealand

The government of Norway and New Zealand, while not abdicating their role as the main promoters of information and communication technologies, have each tried to co-ordinate their efforts with those of private companies equipped with up-to-date technology and know-how.

The Norwegian government has pursued a policy of collaboration with the private sector and has privatised some state-owned institutions such as the Government Computer Centre.

In Norway the main responsibility for urban management rests with the municipalities; the law allows them to collect income and property taxes from private citizens and firms.

The municipalities control basic services such as: health care, schools, care for the elderly, city maintenance, water, gas and energy supplies, *etc*. These services are partly provided by private companies; participation by the private sector has increased in recent years in accordance with government policy.

The New Zealand government undertook a review of its environmental policies and administration in 1989. The responsibilities of local government were maintained or extended. The central government responsibilities (town and country planning, water and soil policy) were transferred to the new Ministry for the Environment.

The aim of the reorganisation was to involve central and local government and private organisations more effectively in environment and energy conservation, sharing responsibilities between them and speeding up the introduction of ICT.

Turkey

In Turkey there is no public body, at local or central level, to promote and facilitate the introduction of information and telecommunication technology.

The Turkish economy is still only partly industrialised despite great progress in recent years. Therefore the number of ICT applications is small. It was estimated in 1986 that only 70 (3.6 per cent) of 1 900 municipalities in Turkey were using ICT to improve the quality of their services.

The scarcity of skilled workers and office staff well enough educated to be easily trained to use new technology is another constraint on the introduction of ICT.

The plentiful supply of unskilled workers – the result of the country's birth rate being greater than its growth rate – accounts for the unpopularity of every measure to rationalise public administration by introducing ICT and cutting employment, and therefore for political resistance to speeding up ICT.

Germany

The government of Germany and the Bundesländer governments have financed a large number of pilot projects in recent years with the aim of facilitating the introduction of ICT.

This policy, launched before reunification, was made possible by West Germany's wealth, but now German funds are being absorbed by the former German Democratic Republic and by other Eastern European nations. ICT programmes will suffer a slowdown or even halt in the early stages of reunification but will recover when and if the economic transformation of the former East Germany is achieved.

Assessment

Impact of the new technology

The energy savings obtained or announced in recent years in many world cities depend not only on the new systems of regulation and control that are made possible by information technology, but also on technological innovation in itself and, often, on the co-operation of the public.

Technological innovation and information technology are very closely correlated. In some cases technological innovation has sprung directly from new computer models; but in others

– such as the high-efficiency power stations of northern Europe, or the most recent Japanese heat pumps that are up to twice as efficient as those produced until last year – results depend on the intrinsic characteristics of the new production cycles and the machines that operate them.

Similarly, the protection of the environment is the fruit of technological innovation, information technology, and the co-operation of the public.

Reduction of pollution and improvement of living conditions are achieved by reducing energy consumption, introducing new technological solutions – such as, for example, the conversion of public transport vehicles from diesel to gas, or the adoption of electrically powered vehicles for commercial fleets – and also by reorganising the city, using the remote data processing criteria that are fully described in the other chapters of this book.

Overall, the energy savings that are obtainable can be put at between 10 and 50 per cent of present consumption, depending on the plants and urban conditions to which they relate. The return on investment can be very quick in individual situations (a building, an individual installation, a motor vehicle) or it may take 10-15 years for complex energy systems such as the district heating of part of a town.

Investment for environmental protection, when linked to energy installations, can be assessed in relation to the plant itself, whereas it is difficult to assess investment from an economic standpoint when it refers directly to the environment, such as air quality monitoring networks or public lighting control systems.

These must form part of a more general budget, the budget for the "entire city system" considered as a whole; these costs, which are not directly productive but are designed to ensure the community's wellbeing, must be borne by the community itself.

The remote data processing revolution, which has been brought about by the new information technology deriving from the powerful combination of electronic data processing and telecommunication networks, has both direct and indirect effects on the environment and on energy conservation.

Telematics will make it possible to bring work immediately to man rather than vice versa[17] whilst tele-shopping, tele-banking and satisfaction of the many other needs of everyday life by means of the computer and communication networks will reduce the volume of transport and with it energy consumption and consequent pollution.

All of this will make it possible to remove city congestion and create more relaxed and more tranquil living conditions. At the same time the city centre, freed from the burden of mass productive activity, will become the fulcrum of information and communication activity, and will strengthen its role in controlling production, trade and finance.

The assessment of the impact of all this change is only in its early stages, and certainly cannot be summarised in a few pages. It should be mentioned that the speed of change is such as to make it difficult to control once under way.

Central and local government should make very careful and detailed assessments of the social and economic consequences of the application of the new information technology before looking at the consequences for energy and environment, in order to use the technology right from the start in the manner best suited to local circumstances. Its application can only have great advantages for energy saving and environmental protection.

Limitation on the wider use of information and communication technology in the energy and environment sector

The convergence of computers and telecommunications, using telephone cables, coaxial cables, fibre-optics and, where necessary, satellite links, makes possible new services and functions which can produce energy savings and environmental protection in the ways described above.

Restriction of the wider use of information and communication technology comes both from institutions – for the reasons explained earlier – and from the pre-existing structures, as well as from the public.

The radical effects of the new technology require substantial structural changes. Old systems (and old suppliers) have to be replaced or modified – take for example the new district heating installations or natural gas distribution networks, and the replacement of fuels, electric traction, *etc.* affecting an entire transportation system. This can create resistance, and consequently generate uncertainty among politicians and in public administrations.

Often the public will not collaborate or is downright hostile, and convincing people of the advantages of the new technology is sometimes more difficult than implementing the technology itself.

In this context, the innovations that information technology makes possible are much more influenced by human and organisational aspects – "orgware" as it is called – than by the availability of adequate software and hardware. Wide research leads to the conclusion that the broader use of this technology depends very much more on socio-cultural and political factors than on technological advance.

The consequences of information and communication technology in the field of energy saving and the environment

The new technologies will have a great influence on energy saving and the quality of the environment – both directly and indirectly.

This essay has endeavoured to explain their impact from the two points of view. The medium and long-term consequences are less dependent on primary energy sources, owing to energy savings, to the possibility of using different fuels, and to an appreciable improvement in the environment and living conditions made possible by the decentralisation of cities.

The new tele-services which information technology makes available to the public will, as a result of decentralisation, allow people to find a new relationship with the environment (less pollution, improved quality of life, improved inter-personal relationships).

This will be all the more true if local and central government can prevent the construction on the outskirts of cities of those terrible dormitory districts that could turn the picture into a caricature.

The future evolution of information and communication technology in the field of the environment and energy conservation

The future evolution of information technology will probably lead to a new organisational, technical and economic equilibrium, in which it will play its most effective part.

Limitations on its application in the energy and environmental field are, from the technical standpoint, practically non-existent. Rather, the limitations are economic or organisational in character, or to do with the reliability of the systems.

Over-complex systems, controlled by insufficiently reliable hardware and software, are not useful innovations but harmful ones.

The most immediate future developments will be remote-controlled energy installations capable of obeying software which is designed for specific results (saving, safety *etc.*) and which ensures the safe and reliable operation of equipment: regional telemetric systems for measuring the quality of the air, which can manage the thermal installations of entire climatic areas, optimising their operation; systems for monitoring and controlling environmental installations (urban, hospital and industrial refuse disposal, urban and industrial sewage and waste water treatment); complete traffic control systems. In short, systems for bringing about a revolution in our way of life, based on the collection, processing, storage and transmission of every kind of information, with all the benefits in regard to energy saving and the quality of the environment which have previously been described.

The choice of energy and environmental policies, and recommendations for their implementation

Now is undoubtedly a suitable time to make everyone understand that energy saving is the safest policy – in view of the uncertain scenario in regard to primary energy sources – and that the deterioration of the environment, one of our most precious assets, is liable seriously to compromise our quality of life.

Unfortunately, every initiative to correct this situation is viewed by the public with suspicion, and in the majority of cases the private citizen is very unwilling to tolerate the presence of new installations near his own home.

It is therefore necessary to obtain public approval and this can only spring from a knowledge and understanding of energy and environmental problems, and the safety of the new technical solutions that are proposed.

Interactive programmes, an example of which is given in Annex 5, are one of the most direct ways of producing involvement, and wide use must be made of them. At the same time, energy and environmental exhibitions must be organised which are lively and amusing, without the boring didactic approach characteristic of such initiatives.

The schools, right from the earliest years, must emphasise to our future citizens that the problems are real and concrete, and must create in them an ecological awareness that is objective and far removed from the unattainable ideals of an uncontaminated world.

The main objectives of a correct energy and environmental policy for the supply of energy to big cities are inevitably in the saving and recovery of such energy sources and the appropriate use of fuels.

Heating usage

The use of energy for heating purposes best lends itself to radical measures. To achieve the above objectives it is necessary to have a sole energy operator, one with a profound knowledge of the territory and its needs and the ability to take a single, comprehensive overall view of the complex problems to be solved.

The following solutions emerge from international experience: the combined production of electricity and heat, the adoption of multi-fuel boilers, the use of heat pumps possibly in conjunction with co-generation, the use of fuel obtained from solid urban waste, and the recovery of heat and energy from incineration plants or industrial production cycles.

Multi-fuel boilers equipped with systems for cleaning up the products of combustion, which the most up-to-date technology offers, make it possible to use different fuels – coal, light oil and natural gas – practically at will. Alongside these technologies the direct use of natural gas should be considered in a strategic role. It can be used in unfavourable meteorological conditions, not only in multi-fuel boilers as an alternative to coal or light oil but as a further guarantee of environmental protection in the most polluted areas of the city, which it is not possible or convenient to reach with district heating, to fire boilers in individual buildings.

In this way a reduction is achieved in the consumption of primary sources and hence, proportionally, in the pollution they produce, and there is further abatement of pollution with the use of the most modern purification systems and, finally, diversification in the use of primary sources.

It will in fact be possible to choose fuels not only on the basis of economic considerations but also on the basis of techniques and policies of diversification of the sources of supply.

This is particularly important in countries such as Italy and Japan, where energy dependence on oil exceeds 60 per cent, and where in some cities heating is almost 90 per cent dependent on oil products.

Energy for electrical purposes

The combined production of electricity and heat is potentially an efficient method of energy saving and reducing city pollution.

Fuel cells, testing of which is being concluded in several parts of the world, will tomorrow constitute not only highly efficient electricity generators that are totally non-pollutant – thanks to the electro-chemical process which produces the electricity without any combustion taking place – and are able to be brought quickly into service, with an almost constant output over the whole power range, but also producers of spare heat for district heating networks.

Thermo-electric installations, if equipped with the purification systems that the most up-to-date technology allows, do not constitute a threat to cities, even if they are located in the cities themselves.

Transport

Transportation systems, like heating installations, are widespread consumers of fuel, and generate pollutants that even the most modern technology cannot eliminate.

Buses running on diesel cause particular concern. A study by the World Health Organisation shows that they are one of the sources most responsible for urban pollution and, in particular, for the presence of dioxin in metropolitan areas.

They can be converted to run on natural gas or liquefied oil gas or, where practicable, may be replaced by the trolley bus which, at least in Italy, was abandoned for aesthetic reasons.

In this way the pollution produced by internal combustion engines, which is difficult to control, is transferred to the thermo-electric power station, where it can more easily be purified.

The pollution produced by private traffic in the historic centres of old European cities can only be reduced by keeping the traffic away, which requires that local authorities replace it with an efficient public transport system.

Finally, the commercial fleets – taxis and postal, telephone, gas and water services, *etc.* – can progressively be replaced by electrically powered vehicles as soon as the introduction of the new ferro-nickel and sodium-sulphur batteries allows their effective range to be raised to at least 150 km.

CONCLUSIONS

From the picture outlined above it can be seen that technological innovation and the new information technology – or rather, their combination – can ensure a better future for our cities.

The limitations lie in the economic and psychological factors connected with the new technology, as well as in the political will.

The economic factors will slow down the speed of implementation, but it will be the public's co-operation and the politicians' will that will make progress possible.

Once again it is man, and his participation in the interests, the problems and the life of the community, that will finally determine the outcome.

NOTES AND REFERENCES

1. ENERGIVERKEN GOTEBERG (1982), "The Heating Supply Plan", Gothenburg, Sweden.
2. NATIONAL ENERGY ADMINISTRATION (1987), National Swedish Environment Protection Board, "Energy from Waste", Stockholm, Sweden.
3. GISTEL, "Teleriscaldamento in Svezia", Milano, Italy.
4. SWEDISH DISTRICT HEATING ASSOCIATION (1986), Swedish Trade Council, "District Heating, Clean Heat for Urban Areas", Stockholm, Sweden.
5. UNICHAL (1987), Proceedings from "Congress 77", Berlin, Germany.
6. UNICHAL (1989), "Yearbook 1988", Frankfurt, Germany.
7. UNICHAL (1990), "Yearbook 1990", Frankfurt, Germany.
8. GOTAVERKEN ENERGY (1989), "Condensing Flue Gas Cleaning System with Electric Power and Heat Production", Gothenburg, Sweden.
9. NEW ENERGY AND INDUSTRIAL TECHNOLOGY DEVELOPMENT ORGANISATION (NEDO) (1990), "The Development of Super Heat Pump Energy Accumulation System", Tokyo, Japan.
10. MAGLIOCCHI, L. (1990), "La Telegestione del Risparmio Energetico ed Esempi applicatovi", paper presented to Abano Terme AICARR meeting.
11. DURLAK, J.T. (1990), "Project on Urban Impact of Technological and Socio-Demographic Change", National contribution from Canada to OECD/URBA 2000, Paris, France.
12. PROVINCIA DI MILANO, COMMUNE DI MILANO (1989), "Studi per la Valuazione della Qualita dell'Arianella Provincia di Milano", Aggiornamento al 31 marzo 1989, Milano, Italy.
13. IEFE, (1988), IEFE – Universita' Commerciale L. Bocconi, "Monitoraggio Automatizzato dell'Inquinamento Atmosferico in Lombardia, Milano, Italy.
14. HEPWORTH, M.E. (1990), "Application of Information Technology in Transport", paper presented to OECD/URBA 2000 Seminar on "Cities and New Technologies", Paris, France.
15. STADDER, J. (1989), "Air Quality Management in the Los Angeles Area", Convegno "Energia a Milano: Passato Presente e Futuro", Milano, Italy.
16. BITETTO, V. (1989), "Il Trasporto Elettrico nelle Grande Citta", Convegno "Energia a Milano: Passato Presente e Futuro", Milano, Italy.
17. QVORTRUP, L. (1990), "Telework: Vision", OECD/URBA 2000 Seminar on "Cities and New Technologies", Paris, France.

INTELLIGENT BUILDINGS

The tele-management system of the Azienda Municipalizzata Case Popolari e Servizi (AMCPS) of Vicenza (Italy)

There were four main objectives to be achieved:

- a saving on fuel consumption and the management costs;
- efficiency of ordinary and special maintenance;
- the availability of systematic, full documentation on the various installations, so as not to have to depend on the personal knowledge of individual technical experts;
- reduction of the amount of pollution, thanks to a reduction in the consumption of fuel and the improvement of the quality of combustion, as well as the creation of a city network for monitoring the quality of the air.

The tele-control system adopted is very similar to that described earlier.

The end data relating to the 88/89 and 89/90 operating seasons can be summarised as follows:

1988/89 season

Central installations using liquid fuel:

– Average variation in the quantity of fuel burnt, compared with the 87/88 season	– 24%
– Maximum variation in an installation that has had the benefit of plant modification	– 37%
– Maximum variation in an unmodified plant	– 33%

Gas installations:

– Average variation	– 30.68%
– Maximum variation	– 39.82%

1989/90 season

Installations with liquid fuel:

– The installations that had already in the previous year had the benefit of the tele-monitoring service showed a further decrease in consumption of about	– 10%

The data for the "new" installations show the following results:

– Average variation		– 22.84%
– Maximum variation:	in a modified installation	– 35.50%
	in an unmodified installation	– 31.93%

Gas installations:

–   Average variation compared with 88/89 in the case of installations
    that were already previously monitored   – 0.13%

It should be pointed out that the two seasons 88/89 and 89/90 are entirely comparable from the climatic standpoint. The initial investment was amortized in three years.

Contact person: geom. Trevisan – Telephone 0039/444/961500 – AMCPS, Vicenza.

TRY BEING THE MAYOR

The player has at his disposal a certain budget (in proportion to the characteristics of each of the cities of which he is going to try being the mayor) and he can decide to use the available budget for investment or public works that he can select from a menu divided into the following sectors: traffic, heating and public works in industry and the service sector.

The programme shows in real time the effects produced by whatever action he takes, using three moveable indices relating to the following:

- extent of the pollution;
- cost of providing energy for the city;
- social costs (the expenditure that the city population must provide the funds for, as a result of the decisions taken).

There are two other parameters that are not monitored, but which are calculated so as to be taken into account when the final verdict is made on what kind of a job the mayor has done, namely:

- the number of jobs lost, if any;
- the increase or decrease in electricity consumption (with the resultant transfer, in case of increase, of pollution to the places where the electricity is produced).

The game ends when the available budget is exhausted, and an overall political assessment is made of the mayor's decisions.

The programme has been developed by Azienda Energetica of Milan on behalf of the Ministry of Environment.

Contact person: SCS – Via Teullie', 1, Milano – Tel. 39/2/89401195, Fax 39/2/8373379).

Computerised interactive simulation programme for the river Lambro anti-pollution plan

The Lambro, which runs through the suburbs of Milan, is one of the most highly polluted rivers in Italy, because in its short course it crosses a higly concentrated industrial and residential area, where there are also harmful by-products of agricultural production.

The reclamation of the river Lambro is estimated to cost over 5 000 billion lire; it will take many years, and will involve an area used by 5 million inhabitants, including the province of Milano.

The simulation programme also has a "time" dimension and illustrates how the river's condition will progressively deteriorate (disappearance of flora and fauna, poisoning of the water, *etc.*) unless the necessary measures are adopted. The programme asks the player to adopt measures, showing the positive and negative results of the player's decision, as well as the compatibilities and constraints. The sources of pollution are divided into urban, industrial and agricultural examples of different dimensions.

The programme also simulates malfunctioning of the purification plant (the player connects it to a higer load than its rated loads; a load of specific poisons and/or heavy metals destroys the microbe population of the purification plant; the plant breaks down for lack of maintenance). The game is then divided into five successive segments, simulating the same number of river sections. Each of these segments is allocated a share of pollutants, which is small in the first segment and increases in the subsequent ones.

The pollution situation in the river is continuously monitored and made available to the player.

There are in addition random variables that can act either for or against the player (non-pollutant factory is installed; an ecological disaster takes place such as the clandestine discharge of a tanker of toxic waste into the Lambro; a purification plant breaks down because of detective management or because of the clandestine discharge of poisons by a factory).

Programme developed for the Ministry of Environment.

Contact person: SCS – Via Teullie' 1, Milano – Tel. 39/2/89401195, Fax 39/2/8373379.

Computer game on the defence of the environment, in the Gulf of Naples region

This interactive video game presents the situation as regards the pollution of the Gulf of Naples, and stimulates the player to tackle the most characteristic questions relating to the protection of the environment, with reference both to the specific situation and to more general subjects. The player has to guide a symbolic personage (in this particular case, Pulcinella, the popular comic figure beloved of Napolitans) through all of the province, looking for sources of pollution; he has to answer questions on environmental matters, and has to defeat the most tenacious polluters.

The game has distinct parts:
- Monitoring the environment: descriptive maps have been constructed that show the pollution situation at every point in the province.
- The game: Pulcinella must traverse the territory looking for the ''seven polluters'' (seven imaginary personages that symbolise the main types of pollution, namely: solid urban waste, sewage, air pollution, sound pollution, toxic and hazardous waste, dereliction of the land and bad land management).
- Before leaving each zone into which the territory has been divided up, Pulcinella must fight a duel (in a small puppet theatre) with the last and worst polluter in that zone. If he is victorious, he will also win all the points associated with that polluter.

Programme developed for the Ministry of Environment.

Contact person: SCS – Via Teullie' 1, Milano – Tel. 39/2/89401195, Fax 39/2/8373379.

American experience

Very widespread in the United States, on the subject of energy conservation are the programmes for simulating energy consumption, which enable the user (young people, consumers of all kinds, *etc.*) to construct an ideal energy bill based on the user's own effective needs, but using the energy in a rational and optimised manner, which makes it possible, by comparing the ''ideal'' bill with his own actual bills, to assess how much lack of rationality there is, and where it is most lacking, in the user's own energy habits.

Programmes of this type are now used in the largest US utilities; they form part of public involvement campaigns, namely those carried out by APPA (American Public Power Association) and distributed to the associated utilities.

Examples of more generalised environmental and energy information, for involving vast masses of population in the great public centres, are frequent in many cities in the United States.

In New York, in the centre of the city (the Chrysler Building), Con Edison has set up a permanent exhibition covering $600m^2$ on the subject of energy saving, and at the exhibition great use is made both of computerised simulation programmes and of other systems of communication based on the use of information technology.

Part 3

MAKING GOOD USE
OF NEW TECHNOLOGIES IN CITIES:
FACTS, REFLECTIONS, RECOMMENDATIONS

A. Testimonies by Cities

GREATER PARIS AND THE WORLD'S METROPOLISES

Michel GIRAUD
President of the Association of French Mayors
President of METROPOLIS, the World Association of Great Metropolises
FRANCE

We used to think of city organisation in terms of physical constraints, but advancing technology is prompting us to design the structures of community life around poles of communication that may take a wide variety of increasingly sophisticated forms, both tangible and intangible.

New communications technologies naturally thrive best in a major conurbation, but it is worth drawing attention to the advantages they can also offer in developing rural areas.

New communications technologies have two functions:

- in improving regional integration, by breaking down institutional and social barriers and pushing back frontiers;
- in helping to strengthen the international role of the major conurbations, by simultaneously promoting competition and complementarity.

New communications technologies as instruments for strengthening regional unity

These technologies are opening up urban culture to increasing numbers of people. Example: the Minitel in the Ile-de-France, an ideal instrument for generalising access to information.

In 1984, the Ile-de-France Regional Council pioneered the introduction of Minitel by encouraging the distribution of terminals which offered wide access to information services using a simple language that everyone could understand.

By late 1989, the region had 1 250 000 Minitels in service. Teletel traffic (excluding directory enquiries) had reached 21 million hours communication.

The new communications technologies are strengthening partnership at regional level.

Most of the leading teleport programmes started and planned throughout the world, especially for New York, Tokyo, Rotterdam and Amsterdam, have been local authority initiatives in partnership with other public sector and private bodies.

In Ile-de-France, too, the Regional Council has set up a Teleport Study Association, bringing together public sector and private sector agencies such as EDF-GDF (electricity and gas), the Caissse des Dépôts et Consignations and the Compagnie générale des Eaux.

But in the context of this conference we could single out the Cologne Mediapark, which includes a teleport, on which construction began last March. This involves the City of Cologne and the Land of Westphalia, together with public sector and private firms, including the Bundespost, which is making it a pilot project.

The project is intended to produce a new urban concept. The Mediapark will be a centre for new media, providing performers, writers, composers *etc.* with the most flexible and efficient communications techniques.

This it seems to me is an example of co-operation not only with a view to social and cultural integration at urban level but also on a European scale since Cologne ultimately aims at becoming one of Europe's leading media centres.

Implementing the new communications technologies is strengthening exchanges between conurbations

As we have seen, the new communications technologies, though usually introduced as a regional partnership, are leading to international co-operation.

Here are some examples:
- the Ile-de-France Teleport Association holds regular discussions with the Cologne "Mediapark";
- EUROSPACE, the European advanced continuous training programme, is involving European co-operation.

EUROSPACE was introduced to serve Europe's industry, universities and research centres with satellite transmission to provide wide coverage of courses and special events.

The plan to connect up Europe's leading museums, being conducted at the initiative of Berlin and partly financed by the EEC, will enable the visitor to any museum connected to the system to glimpse the riches of them all by consulting a multimedia terminal.

These arrangements can however involve difficulties, especially with incompatible technical standards. These have had to be cleared up by agreements on such matters as choice of standard, common hardware, and storage methods for the vast quantities of data required.

This is a field in which the leaders of our major conurbations must be expected to contribute effective assistance.

Technical co-operation of this kind can be based on agreements and understandings between the conurbations.

For example, the regional governments of Tokyo and Ile-de-France have agreed to co-operate on feasibility studies for a distance translation centre between Paris and Tokyo.

Similarly, Ile-de-France and Montreal have agreed that the Ile-de-France Teleport Association should investigate the scope for co-operating with Montreal teleport on graphics.

I hope, too, that on the occasion of our next Triennial Congress to be held in Montreal in 1993, we shall be able to arrange for simultaneous transmission of the discussions by videoconferencing from the teleports of Montreal and Paris. Ile-de-France residents will be able to participate in the assembly, ten years after the creation of Metropolis in Paris.

In today's shrinking world we can more and more easily move about, trade and communicate with one another, as new technologies give us power over time, over space, even over what people do and how their minds work. What impact all this will have on future generations is still hard to assess.

We must learn not to play sorcerer's apprentice with these technologies, but must make sure we keep them to a human scale for the benefit of man.

RECONCILING NEW TECHNOLOGIES, HUMANISM AND DEMOCRACY

Antonis TRITSIS
Mayor of Athens
GREECE

The city of Athens embodies a remarkably rich historical and philosophical past. It is where the idea of democracy was born, two thousand years ago, and even today, in modern Greek, to be the Mayor of Athens means, to preside over the assembly of all the Athenians. The ancient Greeks also forged the concept of the "polis", the city; it is a concept of equilibrium between man and man, between man and his social environment, his physical environment and his institutional environment. This philosophy continues to underlie the goals which the new team of city officials has set. Their goals are: to rediscover the historical physiognomy of Athens, for it is a heritage that belongs to the entire world; to instill new life into the democratic process, enabling citizens to be fully informed and thus restoring their sense of membership in a community; keeping Athens on a human scale by fostering the renascence of the city's 129 districts.

At the same time Athens must become a modern and functional city. The levels of decision-making power in the municipal administration – which has no fewer than 9 000 officials – must be centralised and given sufficient autonomy. A tramway must be put in operation, so as to improve the transport system. Environmental and pollution problems – which are actually less severe than they are often said to be – must be solved through measures to help firms modernise their operations. That is another vital part of the overall modernisation programme; it presupposes co-operation between enterprises and the university, which must become an "open university".

Lastly, every effort must be made to ensure that Athens will not only continue to be an economic pole of attraction but will in fact become a full-fledged international financial centre.

Information and communication technologies can be very instrumental in achieving these goals. But they must be used wisely if they are to become a genuine instrument for human liberation. Things change, cities change, but certain eternal verities do not change: the biosystem does not change, nor do basic feelings, nor the striving for dignity and love. All that is essential to our age-old universe does not change. There are spheres; there is the communications sphere; and ICT networks must be at the service of those spheres. Only if we achieve that will we be able to progress.

PROMOTING AN EFFICIENT CITY,
OPEN TO THE WORLD AND TO ITS INHABITANTS

Jordi BORJA
Deputy Mayor of Barcelona
Vice-President of the Urban Community of Barcelona
SPAIN

Barcelona's modernisation policy, launched in the past ten years, relies extensively on information and communication technologies. This is an important advantage in three fields: promoting the city, modernising its management, integrating its inhabitants and making its population coherent.

Modernising the city's management: the challenge facing the city was to provide more and better services through a territorial and functional decentralisation and through re-organising human resources by providing the staff with adequate technical logistics. Staff expenditures rose by 100 per cent between 1980 and 1990 but as a percentage of the city budget they declined from 50 per cent to 37 per cent. A computerisation effort ($7 million a year of investment and $17 million a year of upkeep and updating) made it possible to achieve a rate of one keyboard and screen for 1.8 agents (technicians and medium level executives). But modernisation does not depend on computerisation alone. For several years, Barcelona has been determined to make massive and permanent use of the telephone to communicate with its citizens. By launching the ''010'' system of 24-hour-a-day telephonic questions and answers, which is extensively used, the city has taken a giant step in that direction. Direct lines are also used so that citizens can ask their questions and make their requests in such fields as public works, cleanliness and traffic. Still another service enables people to take care of their administrative formalities by telephone. The effort is being continued today with the creation of systems for telematic consultation and single ''tellers'' so that the city's inhabitants can go about their administrative business without being hindered by separations between departments.

Promoting the city: Barcelona's aim is to be the pivotal point between northern and southern Europe. This requires both reactivating the economy and revitalising the urban fabric. Public and private co-operation for such institutional actions as ''Barcelona posa-t-guapa'', promotion of general interest investments (''Initiatives, SA'') and the elaboration of the ''Barcelona 2000'' Strategic Plan are examples of ways in which resources are mobilised to this end. Furthermore, if the conurbation of Barcelona is to be truly competitive, it must be accessible, and must have a strong communicatiions infrastructure not only in terms of physical networks but also in terms of telecommunications networks and systems.

Making the population integrated and coherent: the city cannot be open to the outside world unless it is capable, in the context of its own operations, both of providing its inhabitants with the information they expect and of receiving the information that they wish to channel to it. Concerning information and communication with Barcelona's citizens, it is significant that territorial decentralisation also made it possible to experiment with new forms of participation: citizens' participation in the management of infrastructures and public facilities, the citizens' right to make proposals concerning regulations. Decentralisation has been crucial in revitalising the most deprived neighbourhoods. This policy presupposes a change in regulations which is expressed in the city's Municipal Charter. Finding the legal and regulatory innovations needed to move ahead in this direction will take careful thought.

Administrative streamlining will depend on the political aims of the elected officials. The productivity of services provided must increase the social cohesion and wellbeing of the inhabitants who will benefit, at the same time, from the economic promotion of the city at the European level.

The use of new ICT presupposes not only a change in organisational models but, often, a change in local policies as well. More and more frequently, the local authorities must set up contractual mechanisms with the State authorities and with private firms.

Cities are places where the policies designed to solve the major problems of our day can be tried out. One aim is to find equilibria between growth and environment, mobility and security, privatisation and the public management of services. Coming up with answers to these questions will require developing ''city networks'' which will be able to inform one another, exchange their experiences with regard to innovation and devise joint policies. This aim should be a major goal of a Europe which is to be based on strong and dynamic urban units as living entities.

TORONTO, THE HEALTHY CITY

Arthur EGGLETON
Mayor of Toronto

Perry KENDALL
Deputy Mayor of Toronto
CANADA

Toronto has long been proud of its environment, its relatively low crime rate, strong multi-cultural neighbourhoods and its vibrant city centre. But we have not been untouched by the problems that beset urban centres throughout the world: housing shortages, poverty, congestion, pollution, urban alienation, drugs and violence.

In the early 1980's in an initial response to these problems, the Mayor's office organised workshops with community groups to discuss how the city could surmount its problems. That was the beginning. Four years later after much consultation, the City Council approved a report called "Healthy Toronto 2000", drawn up by a representative task force, which made 89 recommendations to keep the city in "good shape". This document was unique. Its definition of "healthy" included the broad range of environmental, social and mental factors that determine the welfare of those who live in the city. Sub-groups focused on managing growth, improving the quality of life and promoting social equity. Throughout there was extensive city-wide consultation with public meetings, surveys and in-depth discussions with hard-to-reach groups such as the "homeless".

Like many North American cities, Toronto has historically under-utilised one of its most precious resources – land. Nowhere is that more evident than on our main streets, where miles and miles of main roads are flanked by one- and two-story buildings, mostly commercial properties. This has led to very low densities on these central arteries and to urban sprawl. The city is now looking at ways to encourage private developers to build four or five stories of housing over one-story retail stores on shopping streets. These houses will provide much-needed housing for under-served groups such as young single people, childless couples, seniors and single parent families.

Toronto is also facing up to four of the most intractable urban issues of the late twentieth century – homelessness, poverty, crime and drugs. And we are trying to do that in innovative, human ways. "Street City", a non-profit development of 72 individual units in an unused warehouse owned by the City, caters for some of the hardest-to-house men and women living on the streets, those who have been rejected by ordinary homeless shelters. They worked side by side with master carpenters, transforming the warehouse into their own community, learning new skills in the process. If the project proves successful, it will be a model for other initiatives to solve homelessness.

Issues of crime and security, particularly their effect on women and children, are being tackled by our Safe Cities Committee, the first of its kind in North America. The scourge of drugs has also prompted action through the Mayor's task force on drugs. This group is co-ordinating intensive prevention, enforcement and treatment effects throughout the city. Similar task forces have been established to deal with other issues such as hunger, race relations and ageing.

The Healthy City Office will continue to monitor progress on these issues by measuring changes in the city and its progress towards its goals every three years. A key target is improvement of the environment. Toronto has committed itself to a 20 per cent reduction in carbon dioxide emissions by the year 2005. A number of major steps are being taken to achieve that goal, including a review of ways to reduce automobile pollution, the setting up of an energy efficiency office and an ambitious city-wide tree planting programme. Toronto already has a large-scale residential recycling programme for metals, glass, newspapers and plastic. And the City Council has approved an initiative that will require fast food restaurants to provide re-usable cutlery, dishes and other tableware. The potential impact of this is substantial since some 200 million disposable containers are thrown away in downtown Toronto every year.

If environmental protection, fairness and security for citizens are important goals, the bottom line is not what a particular project will cost but what it does to preserve the quality of city life. That is a truly representative bottom line. That is what "building a healthy city" is all about.

TELECOMMUNICATION AS A COMPETITIVE STRATEGY IN AMSTERDAM

Jaap ENGEL
Director of TCA-ARIT, Amsterdam
Foundation for Information Technology
NETHERLANDS

Amsterdam regards the introduction of modern information systems and telecommunications as part of an overall competitive strategy to increase the attractiveness of the city as a business centre. Among developments the City Council is helping to introduce are:

- telecommunications networks;
- cable television;
- high technology industrial developments;
- a World Trade Centre; and
- an interlinked cargo system at Schiphol Airport.

Telecommunications: the Dutch PTT has established a fibre optic network connecting the main centres in the Amsterdam Metropolitan area.

Industrial developments: telecommunications, technology and information businesses are being concentrated in three main areas: a commercial cluster around Teleport Sloterdijk jointly developed by the City Council, Dutch PTT and private interests; Informatics House, near the Central Station, for use by high technology firms; a new science park, Watergraafsmeer, for specialist research into chips, solar cells and artificial intelligence.

The Amsterdam World Trade Centre: this has 74 320m^2 of office space with a video conference centre provided by the PTT. There is also a demonstration centre for advanced telecom services.

Schiphol Airport: in addition to offering an integrated telecommunications service, the Airport Authority is developing a value added network (VAN) to link air cargo carriers such as KLM, forwarders, brokers, truckers, the railways, shippers and Dutch Customs.

Another project under development is a *road-traffic guidance system.* Strategically placed signal points along city roads will give information to on-board computers mounted in cars and lorries as a way of solving traffic congestion problems.

Overall the role of the Dutch PTT and other telecommunications providers is critical to the success of Amsterdam's strategy. The services offered, the quality and pricing of telecommunications, services for voice and for data, for local, national and international communications, are all major issues. That is why Amsterdam aims to work continually and constructively with the telecommunication companies to enable them to achieve at least a comparable competitive position with rival European firms and to identify and promote new services. Our aim is to create an "intelligent" Amsterdam.

OSAKA, AN INFORMATION-CENTRED CITY

Toshiaki NINOMIYA
Deputy Mayor of Osaka
JAPAN

In Osaka we are aiming to create an urban environment for the twenty-first century. A series of "intelligent" buildings are currently under construction complete with the latest telecommunication and information systems. The core of the "intelligent city" plan is Techno-port Osaka, a large scale project which will encompass the construction of three man-made islands in Osaka Bay. These islands will play a major role in the growth of the Kansai region as an international metropolitan centre designed to permit round-the clock exchanges of goods, information and people on a global scale.

One of the key facilities, Intex Osaka, an international exhibition centre and an important base for international trade, was opened in May 1988. Three months earlier, System House, a research complex for smaller companies in the fields of advanced electronics, was inaugurated. Other facilities currently under construction are a World Trade Centre Building and the Asia and Pacific Trade Centre. A key facility still being built is Osaka Teleport, an earth station for satellite communications. According to the present plan, the teleport will comprise four parabolic antennas around a centre block incorporating transmitters, power units and other facilities. The earth station will enable direct communication between Osaka and cities in America and South East Asia via a telecommunications satellite above the Pacific Ocean. Communication between Osaka and other cities throughout Japan will be made via another satellite centred above the south of Japan.

An information communications network based on Osaka Teleport is being developed. Already an optical fibre network of 1 800 km has been completed linking 133 cities, towns and villages in six prefectures in the Kansai region.

All this is designed to promote Osaka as a focus for information and communication. But we believe that in order to create "an intelligent city", two further objectives have to be attained. First, city automation systems must be developed for providing such services as traffic information and parking. Second, it is important to create the right environment so that contact can be developed and sustained between citizens and nature and people and culture. That way an environment focused on human needs can be provided for in an "intelligent city" ideally suited to the requirements of the coming information age.

MINATO MIRAI 21

Yasuhei SATO
Deputy Mayor of Yokohama City
JAPAN

Rapid evolution of the information society is requiring qualitative adjustments to the structure and functions of cities. In addition to the functions already supporting urban life, it has become essential for cities to have the facility to readily produce, process, circulate and consume information for business use and daily life. Such access to wide-ranging information, in turn, brings flexibilty, creativity and vitality.

The advantages of information-oriented urban functions are being emphasised in leading urban development projects undertaken around the world. Domestically, in cities such as Tokyo, Osada and Kobe, the movement promoting urban development using state-of-the-art telecommunications technology is strengthening. National support for the development of autonomous cities has made possible further development of various information-oriented policies through the provision of grants, interest-free loans and favourable tax treatment. The Japanese Ministries of Construction, International Trade and Industry, and Posts and Telecommunications have been particularly active in this regard.

The City of Yokohama is progressing with an urban development project that will incorporate telecommuncations technology utilising the support provided by these national agencies. The project, known as ''The Minato Mirai 21 Plan'', is a mojor urban development undertaking which aims to reform the waterfront and strengthen the city's autonomy. This is being achieved by redeveloping the run-down dock area on the city centre's waterfront, and creating an environment able to support a working population of 190 000 and a residential population of 10 000, on a total area of 186 hectares. In striving to realise its vision of an information-oriented city of the twenty-first century, this development has incorporated the ''Minato Mirai 21 Teleport Plan''.

Firstly, a ''Communications Centre'' will be established on the Minato Mirai 21 site, acting as a central facility providing information and communications services. The Communications centre will house Nippon Telegraph and Telephone Corporation's (NTT) Integrated Services Digital Network (ISDN) exchange, a facility with the capacity to handle an enormous volume of communication. The ISDN exchange will be the base for a high-capacity, quick-transmission optical fibre network running through the Minato Mirai 21 district, which will serve as a communication network with other cities. In the field of international communications the inter-exchange carrier company, International Digital Communications Incorporated, has already built an ''International Operations Centre'' in Minato Mirai 21's neighbouring Portside District.

The world's first request-based moving-image information service has been installed in Minato Mirai 21 and in other central areas of Yokohama. This system, known as the Video Response System (VRS), has permitted the provision of a more vivid and effective visual information service than that allowed by videotex through the use of optical fibre cables.

The development of a safe and comfortable city is one of the major issues to be addressed in town development. One means of achieving this using telecommunications technology is through the Public Facility Information and Control System. This System, to be installed at Minato Mirai 21, will gather and control town information, and information on disaster prevention and the operation of facilities. All facilities on Minato Mirai 21 will be linked to the System Centre by a transmission network, which will centrally manage the administration and control of public facilities.

Other information sevices under examination are multi-channelled CATV designed to accommodate high-definition television (HDTV), a data base to support business activities, and the Integrated Circuit (IC) Card; only recently has it been possible to put it to practical use in Japan.

Some final issues remain in regard to the use of telecommunications technology in urban development.

Firstly, an institutional support framework is necessary in order to permit the use of these state-of-the-art technologies. Until only ten years ago in Japan, it was not possible to utilise telecommunications technology in the context of urban development. However, now that the communications sector has been liberalised, there is common ground between urban development and communications, making it possible to undertake communications projects as a part of urban development. One instance of this is a project being undertaken by Osaka City and valued at Y 4 000 000 000. Nevertheless, the legal framework underpinning these types of projects is still inadequate in certain areas.

Secondly, the relationship between people and technology should be closely considered when telecommunications technology, or any other technology, is to be used in urban development. Information-oriented cities should provide urban functions which enable citizens and businesses to be creative in their activities and lives. Technology introduced with urban development should not simply be used with the aim of exploring technological possibilities, but must also play a positive role in the lives and activities of people. In this way, the effective use of information-oriented technology in urban development will not only advance our quality of life, but will also assure us the development of modern, vital cities.

URBAN PLANNING IN KOSHIGAYA

Shinichiro SHIMAMURA
Mayor of Koshigaya

T. SEKINE
Head of Planning Department
Koshigaya
JAPAN

Advanced information systems will be crucial in controlling the future of all cities. In Koshigaya, with its population of 284 000 located 25 miles north of Tokyo, we have devised a plan which we call "The Intelligent City". It is focused on three priority districts within the city and aims at realising within them a highly advanced information centre which is autonomous, multi-functional and densely populated. In addition, the plan aims to coordinate other big projects such as the construction of an elevated railway, the development of a new town and the construction of an expressway for cars. With this plan, Koshigaya hopes to realise the citizens' needs for security, convenience and amenity in a balanced, high quality urban life in which the work place, home and recreation facilities are all close to each other.

A key part of this endeavour are information, communication and software systems. We are planning to introduce computerised traffic information and guidance systems in addition to improving the bus service. We are building support tele-communications for business, research, distribution, and satellite cities. A disaster prevention information system is being created for greater security. Already a "hot-line" connects offices such as the Tokyo Electric Power Company, the Nippon Telegraph and Telephone company, the fire department and the waterworks divisions. Radio links are also set up in the city to provide accurate information quickly to citizens in case of emergency. A river control system has been established in case of a typhoon or sudden heavy rainfall enabling flood watergates to be immediately controlled.

Koshigaya is now the 69th largest city in Japan. It has made and is continuing to make steady progress. Our city aims to provide security, convenience, and amenity – in short a high-quality urban life.

TOWNS AND NEW TECHNOLOGY IN BADEN-WÜRTTEMBERG

Werner HAUSER
Chief Executive Officer, Association of Local Authorities
of the State of Baden-Württemberg
GERMANY

The Data Central Office (DCO) handles the development of electronic data processing systems for communities, districts, and adminstrative offices in Baden-Württemberg. It advises all public data processing users in the state and is also the centre for training and education of public employees in electronic data processing. Under the DCO, seven regional computer centres link 90 per cent of the state's adminstrative offices.

This highly centralised computer network has brought the use of information technology to a high level within the state. A tight and efficient infrastructure for information processing has been set up in the last 20 years while at the same time the importance of muncipal self-government has been maintained.

The DCO and the municipal computer centres see themselves as public service enterprises competing with private enterprise. Each of the 165 towns within the state are free to choose which service they want. The DCO offers standardised up-to-date software for a variety of municipal tasks using the latest software technology including:

– modern design of transaction-oriented programmes;
– standardised use-oriented interface;
– short response times;
– professional safeguards for data protection;
– state of the art software development tools;
– high integration of application programmes;
– training and demonstration centres.

The main applications of information technology in Baden-Württemberg include car registration, personnel, pay roll, management of citizen's registration. Other developments in progress will cover an environment information system, digital maps, graphic data processing and the processing of property data.

Today's advanced communal data processing organisation stems from the view that only cities and communities working together right across a state can use modern data processing facilities in an economical and practical way. One of the main tasks for the future will be to make the transition from centralised data processing to smaller office operations.

COMPUTERISED MANAGEMENT OF A "CITY ENTERPRISE" TO IMPROVE THE FUNCTIONING OF THE "CITY AS AN ENTERPRISE"

Gianfranco DIOGUARDI
President of Bari Technopolis
ITALY

Summary

The Bari Technopolis is a consortium comprising the University of Bari (a town in southern Italy), public bodies and private enterprise. Much as with similar bodies in France, its task is to carry out practical experiments based on the findings of applied research and innovatory processes developed within the consortium.

For instance, the Technopolis promotes and supports general and specialised training activities along with vocational and technical retraining, all at local firms, which as a result can integrate any innovation more smoothly and effectively.

The Technopolis is in the process of developing an innovation and training project directly aimed at the city, which is treated as an entity operating like an enterprise (the town of Bari has been selected as the current object of the experiment). With this aim in mind, and in collaboration with Dioguardi S.P.A., a construction firm in the consortium, plans are underway to establish programmed maintenance for the "enterprise town".

By means of computerised management, the project supplies maintenance programmes for public and private buildings and infrastructure in the town. Interaction is also planned between smart terminals, installed throughout the city (the "district labs"), and the residents, in other words the users of the housing and infrastructure covered by the maintenance programme.

Interaction with local residents is not confined to the actual upkeep of existing installations, but also provides a new social and cultural concept of what social activities, or "social development" should be. For instance sporting activities seek to involve underprivileged youth from the suburbs. Local interaction of this kind can help to bring about a "united city", thanks to telematic links between various "district labs" in the suburbs. Participation in the project by FORMEZ, a government body dealing specifically with the area of Southern Italy ("Mezzogiorno"), focuses on training issues.

In short, the Technopolis, a construction firm and a government training body are all working together to carry out a "maintenance" programme for the suburbs, not only to ensure the upkeep of urban facilities but also to maintain the city's social and cultural fabric, the ultimate aim being to unify them and make the term "united city" a reality.

MULTIFUNCTION POLIS

Paul BARRATT
Special Adviser, Australian Department of Foreign Affairs and Trade
AUSTRALIA

The Australian and Japanese governments are jointly planning a "city of the future" to meet the technological, social and economic challenges of the twenty-first century. In addition to being a world class centre for advanced research and industrial development, the design of the Multifunction Polis (MFP) is aimed to provide an ideal environment for city living. A joint steering committee of seven Australian and seven Japanese members was appointed by the two governments to provide high level advice and guidance and a feasibility study was begun in November 1988.

The pilot concept, developed for the MFP, is centred on three key elements:

- a "technopolis" featuring information and communications, advanced transport services, and construction and design;
- a "biosphere" in which the city of the future would become a major centre in the international environment management industry;
- a "renaissance" living style in which all citizens' needs were specifically catered for health, education, leisure and media.

The basic concept proposed by the Japanese Trade Department (MITI) assumed a population of 50 000 to 100 000. A site was selected in Adelaide, the capital of South Australia, after exhaustive competition between the different Australian states. The site proposed by the South Australian Government is a greenfield development of over 3 500 hectares only 20 minutes from the centre of Adelaide. The aim is to seek to retain the economic, social and cultural advantages of higher density cities while not suffering from the crowding, pollution, anonymity and alienation of many large existing cities. To achieve this the South Australian proposal involves the creation of a "city of villages". In the words of its proponents the design concept for MFP-Adelaide will:

- re-establish the relationship between those who live in cities and the natural environment surrounding them;
- recapture the sense of community; each village will be a cohesive unit managed by its own council;
- provide an ideal environment for development of a technology-based economy;
- welcome the integration of new residents.

A critical element in the design concept is new technology. Information systems and new forms of transport and communications will enable the efficiencies of big cities to be retained in a smaller community. Industry and commerce will operate at various levels without losing that human scale essential to a high quality of urban life.

The MFP will use state of the art technology and design to deliver to its citizens a lifestyle and standard of services which offer the best that twenty-first century living has to offer. In so doing, it must operate in an environmentally responsible way and provide an appropriate role model for the development of new urban centres elsewhere.

The aim is that the MFP should be planned from its inception as a major exercise in international co-operation, seeking to harmonise the interests and objectives of two very different peoples and cultures, and to do this in a way which is attractive to the international community generally. It will be far more than a new town. From the Australian perspective the driving force is a vision of the contribution which MFP-Adelaide can make to the evolution of the Australian economy and Australia's role in the world as we enter the twenty-first century. It is different from traditional technology park-type projects because it seeks to take a holistic approach. Fundamental to the concept is the integration of a striking urban concept into the framework of the existing metropolitan area so that each enhances and influences the future development of the other. Thus the whole of Adelaide, not just the development site, will become the MFP.

B. Assessments

NEW TECHNOLOGIES, PARTICIPATION, INTEGRATION AND LIFESTYLE

Peter HALL
Professor at the Institute of Urban and Regional Development,
University of California-Berkeley
UNITED STATES

The challenge of the informational society

In his new book *The Informational City,* Manuel Castells argues that we are witnessing the second great economic transformation of modern times. The first was the transition not from feudalism to capitalism, but more fundamentally the change from an agrarian to an industrial mode of production. Similarly, we are now seeing a transition from that mode of production to an informational mode, in which the fundamental inputs are no longer material, but knowledge-based (Castells 1989).

The new informational mode is not the product of new technologies, nor are the technologies a mechanical response to the demands of the new organisational system. Rather, it is the convergence between these two processes that changes the technical relationships of production, giving rise to a new mode of development.

The point about Castells' analysis is that information becomes the basic medium of the new mode of development, just as land was the medium for the agrarian age, or coal and iron for the industrial age. But, unlike those earlier media, it is not fixed in space at all: it can be moved, used, replicated at will, almost everywhere. Potentially, in a way that was never before true, it makes all places equal. Yet nearly all the analysis so far made on this subject suggests that in fact, so far, just the opposite is true: there is a sharp distinction between information-rich and information-poor regions, and it seems to be becoming more extreme. There is a process of circular and cumulative causation, whereby the information-rich cities and regions create more demand for information, which in turn produces more supply in the form of innovative individuals and establishments; and these in turn generate new forms of demand. Just the opposite happens to the information-poor regions. The comfortable notion that weightless information will equalise spatial opportunities, then, may prove a cruel myth.

The shift, so Castells argues, has profound consequences for the rise and fall of cities and regions: there is a "new industrial space", characterised by a sharp spatial division of labor, the generation of information in innovative milieux, the decentralisation of different production functions, and extreme flexibility in location. At the core of the system are the great global command and control centres, remote and apparently outside anyone's control. They locate in a few world cities, which increasingly relate more to each other than to the rest of the world: London, New York, Tokyo. These in turn sit in the centres of vast and spreading metropolitan regions, within which more routine informational functions are constantly banished to more peripheral low-cost suburban locations. Even more routine functions, like service delivery and information distribution and retrieval, can be standardized and spread across the entire national and international space, driven by the new information technologies. These technologies thus permit the spread of control of the leading centres, and cement their power.

We can extend Castells' argument: the same is equally true of individuals. Quite apart from the fact that they may be helped or handicapped by accident, by virtue of where they happen to live and work, some individuals within any region or city may prove to enjoy a far richer access to information than others. Knowledge, of course, was always power; but, in earlier eras, a certain amount of essential knowledge – of how to cultivate or keep flocks, of how to practice basic industrial skills – was widely diffused. Now, however, knowledge through information is becoming progressively richer, more varied and more complex. What is essential is the information about how to acquire information. This is partially acquired in basic schooling, but even more so in the course of the working life. The nature of that working life, including the fact of whether there is a job, acts as the basic filter and controller of information. At one extreme, the long-term jobless worker has little chance of another job, because of lack of information about opportunities. Even access to the so-called informal economy is controlled in that way. At the other extreme, the professional worker sits in the midst of an entire network of informational contacts, both formal and informal.

New technologies, which now drive the development of the economy, unfortunately also create an increasing gap between two classes: the information-rich and the information-poor. It is perhaps no exaggeration to term this the new class division of the late twentieth century, replacing the traditional segmentation between capital and labour. A whole class of society, living in what the media call the new urban barbarism, are excluded from effective access to information and so from participation in the mainstream economy and mainstream society, which define themselves in terms of the use of information. Small wonder, then, that massive alienation is the result.

And, even for those who succeed in finding a niche in the middle levels of the informational economy, the resulting work – relating to machines rather than to people – may itself be deeply alienating. Telework offers the promise of a return to the integrated living and working patterns of the pre-industrial age, but even it brings stresses of its own: in particular, the need to combine two kinds of role, the professional and the familial, at one and the same time.

Lessons from the papers

It is not too fanciful, here, to speak of a right of access to information. As Gabriel Dupuy's paper reminds us, many advanced countries guarantee citizens a certain minimal access to networks such as telephone or electricity; adequate basic services are now not a luxury, but a basic norm. But the same author notes that informational networks are still relatively primitive, resembling the automobile network in the early stages of its evolution. Yet, he thinks, one can see the signs of development of such a system. He cites the development of integrated systems to control and charge for other urban services, now operating at an experimental level in France and Germany, and of geographical information systems which provide a complete map of the nature and distribution of such services. However, the critical question is the access to such services. In the most advanced cases, such as Tama outside Tokyo, the public have only a restricted access to the very rich information that is potentially available. However, such systems prove to be relatively cheap to install; cost is rarely a barrier to experimentation.

Gilbert Paquette's paper reminds us just how restricted, over most of human history, access to information has been. The Sumerians invented writing 4 000 years ago, but until 500 years ago the lack of printing meant that there could literally be no broadcasting of the written word. And even during most of this modern era, all the work of acquiring, processing and communicating knowledge was the work of a small minority of intellectuals, professors and writers. All that suddenly changed about 40 years ago, in the transition to the informational society: now, in all advanced countries, a majority of people earn their living by handling information. This in turn has brought enormous demands on the educational system, and a huge increase in demand for educational and cultural services.

Paquette goes further: in the beginnings of this transition, "informatisation" extended only to simple tasks and routine information. Now, we see a "teleportation" of all forms of knowledge, through the use of artificial intelligence and the development of so-called hypermedia which combine all available techniques to acquire and process knowledge. Paquette calls this new stage "distributed cognitive information systems".

He goes on to identify five great "construction sites" for the new informational society: schools, workplaces, home study, "mediatheques" and alternative community services. Schools will be transformed into self-study centres; they will break away from their "Taylorist" model and try to "post-industrialise" education through new ways of apprenticing professionals in the basic competencies they will need in the coming world; no longer, as he puts it, quoting Denis P. Doyle, will the teacher be the worker and the student the product; instead, the teacher is the generator of apprenticeships while the student is the worker. Second, the workplace will become a place where the "Taylorisation" of training gives way to a system in which continuing education is embedded in the processes of work themselves; this will require a joint effort between governments, local authorities and enterprises. Third, learning in the home through educational television and inhome informatics; tele-education by cable will play a vital role here. Fourth, existing museums, libraries and cultural institutions will need to transform themselves into "mediatheques", multi-media centres with a great variety of specialised informational resources: "the second wave of informatics", he tells us, "will facilitate (the) interactive integration of the media in time and space". Finally, side by side with this, there will be a great variety of specialised communal informational resources targeted to sections of the public. And all these different informational services, he stresses, will finally be integrated into a seamless web of information, "a global cognitive process".

Paquette gives us examples of the beginnings of this process in different OECD countries: the introduction of personal computers in the schools, the use of telematics in language education, new systems of professional qualification, distance learning through tele-universities, the French Minitel system and its potential for education, the use of CD-Rom in museums and libraries, municipal "information cottages" in Finland. They are the beginnings of a revolution. But he concludes by warning of limits to their diffusion: the costs of the initial investment; the ignorance about the technology – which leads both to exaggerated claims and exaggerated anxieties – and the problems of making it user-friendly; and the problems of socio-economic and linguistic diversity ("very often", he points out, "the technology talks English"), which may penalise certain social groups, women and older people, even certain nations. These will be overcome: he points to the importance of anti-sexist education in schools, for instance. But it will take time and effort.

Bengt Söderström's paper deals with a specialised but vital aspect of Paquette's grand schema: public information systems to serve neighbourhood needs. He stresses systems intended to give citizens better information about public authorities and organisations, and also opportunities to influence them. These systems would help citizens by providing access to many information systems at one place. Presently, such information is segmented, making it difficult to obtain but also reducing the relationships between people. One result, he writes, is that intended efficiency gains often give rise to many negative social effects... (an) aspect of neighbourhood integration is an effort to restore/vitalise local social effects, to give people better control over their living conditions and to encourage shared responsibility for common, local affairs.

At present, though technical development is very rapid, the tendency is for different organisations to build their own systems, resulting in a very disparate picture: "different types of systems overlap and compete with one another". In particular, public and private information systems have different strategies: the private sector stresses flexibility, freedom of action and tailoring systems to meet its customers' needs, while the public sector stresses co-ordination and co-operation to reduce the costs. Together, these strategies come into conflict and make it more difficult to produce an integrated system.

His examples come mainly from Sweden, and most of them are very specialised: home ordering of pharmaceutical and convenience goods for the less mobile, theatre ticket sales through post offices, and so on. But he also describes a more ambitious experiment: CISP, the Citizen Information System and Citizens' Office, on trial in the city of Härnösand in Västernorrland County since August 1990. A co-operative effort of central government, county and municipality with the Swedish National Telecommunications Authority and the local authorities association and its computer services company, this experiment is first aiming to build a simple information system for groups like pensioners, parents, unemployed young people and home-builders. In some areas the system has already evolved beyond supplying standard boilerplate information, to offering personalised service. At present, and for some time to come, the system will be operated by an official, who will lead the customer through the questions and the answers. Similar in many ways are the service shops being operated in some 15 Danish cities, the citizen office in the city of Unna in northern Germany, or the 31 neighbourhood offices in the city of Walsall in the English Midlands. All these use computer information to some degree, but none perhaps so centrally as the Swedish experiment. The central idea in all these experiments is to create "one-stop shops", which provide not only general public information but also qualified individual advice and can do some kinds of casework. This raises complex issues as to the relationship of this "front office" service and that of the sectoral agencies. The use of computers, and expert systems, can bring the specialist's knowledge to the front, even though the shop does not have access to the individual expert.

As well as supplying information through offices, the new technologies can penetrate into the home. No country can equal France, with its five million Minitel terminals, on this score. But Sweden, where the equivalent service reaches only 16 000 subscribers, intends to extend to 500 000 households by the end of the 1990s. The Swedish strategy is different from the French: while the French distribute the terminals for free and gain revenue from the suppliers of value-added services, the Swedes intend to make a small flat charge plus time charges, with 30 per cent of services provided free of time limits.

Such experiments replace centralised, bureaucratic administrative systems by co-operating, dynamic organisations capable of relating directly with the public. But they do more than that: they can help promote a sense of the local community, thus encouraging human contacts within that community. The Scandinavian experiments show that people do have great interest in their local affairs but that they also expect quality, so that the costs could be large. Söderström, like Paquette, ends by stressing the role of information technology in a number of neighbourhood contexts – libraries, which in many countries are becoming local culture and communications centres – and schools and adult education centres.

The workshop

Reflecting on these thoughts, participants in Workshop A – *New Technologies and Citzenship* – began with a summary statement of the issues from the chair, written well in advance of receipt of the papers. Given that fact, it was remarkable to notice how many of the key issues tended to overlap with the authors' conclusions.

These, the chair suggested, were the key issues:

First, the development of information technology had so far been limited in three important ways:

- The relatively weak diffusion of the personal computer into homes (typically 20-25 per cent of households at maximum, with many of these probably not in regular use).
- The lack of convenient, user-friendly ways of communicating between computers, especially from the home. There have been exceptions, for instance the French PTT Minitel system, but they have been few.
- The limited capacity of computers to store images, particularly moving images.

Second, this situation should however change dramatically during the 1990s, because of three major techno-social developments:

- Multi-media systems will become commercially available, combining computing, informational and entertainment functions. In particular, computing and television imagery will become part of a seamless web.
- Communication between computers will be greatly enhanced by the development of broadband telecommunications, including ISDN.
- Because of these developments, the spread of computers (in the form of multi-media stations, linked to broadband networks) should be greatly accelerated.

Third, these changes should encourage the development of user-friendly software, creating an environment quite unlike the somewhat intimidating face which PCs presently offer to the public.

Fourth, therefore, the new technology presented a unique opportunity for the enhancement of democratic processes through interactive simulations. For instance, citizens could be presented with simulated future environments (alternatives for new suburban development, for a new highway or high-speed rail line) and could express their preferences, including intensities of preference. This would allow the development of "budget pie" games in which citizens recorded their strength of preference through the use of simulated resources. In turn, such choices could be integrated into more complex decision-making processes, wherein citizens could explore possible side-effects. It would even be possible, given suitable safeguards, to develop electronic voting systems in elections.

Fifth, however, the obvious danger was that such technologies would encourage an "information gap" between information-rich and information-poor groups in society, which the former group would exploit to their own advantage. This could distort the democratic system in the direction of further inequity, since the information-rich groups are generally the most affluent.

Sixth, to obviate this, governments would need to consider the development of deliberate policies:

- To speed up the diffusion of the new technologies to all sections of society. The problem here is that this process of diffusion may be governed in large measure by commercial considerations. But commercial developers, too, have an interest in reaching the widest possible markets. The right way might be through specific R&D subsidies.
- To encourage the development of user-friendly systems, easily used by those with little previous knowledge or training. A particular role might be played here by state broadcasting systems, especially those that have had previous experience with these technologies (Open Universities and Colleges, the BBC's Domesday project).
- To use publicity to spread interest and enthusiasm for the new techniques, especially to more deprived groups and areas.

Taking up some of these themes, the workshop started with the proposition that only people, working through the democratic political process, can help integrate society; technology alone cannot repair what technology has helped break. But, intelligently used, it can help.

The place to start, the workshop concluded, was in the schools. For, even in the most information-deprived groups and areas, the young are more comfortable with the new technologies than are their parents or grandparents. They can learn them, and show the way to their elders. But this means that the school itself must change its function and its nature, perhaps more radically than ever before in the 4 000 years of the history of schooling. From a place of instruction, where older people inculcated knowledge into younger ones, it must become a meeting place and a resource place where children – and indeed adults – come together to learn how to learn. The technology – a computer to every student – needs to become the basic resource, which the student uses to acquire and manipulate information. But, we are reminded, computers cannot teach values: they are amoral. Old-style families, old-style schools once passed such values from generation to generation. Now, neither of these social units does that job adequately. Our workshop did not emerge with a clear answer to this crucial dilemma.

The second place to work is in the neighbourhoods: we need in effect to establish the quality of the old village in the midst of our amorphous suburbs. Some cities are already trying to do this, with some success: we heard of Brest's efforts to create a twentieth-century agora, as they call it, through electronic bulletin boards through which the citizens themselves exchange information. Here, as elsewhere, the aim is to develop networks. This may require far more widespread technological diffusion than most OECD countries now possess: computer power and cable TV for every household, as a minimum requirement.

That raises the question: who provides this network, and how? The user will not readily perceive the difference between a public and a private network. But the market is driven by demand, and unless societies become much richer quite quickly, or unless there is massive and unprecedented income redistribution, that could take generations, if the market is left to provide the system. So the public sector may need to intervene, at least to participate; but that raises practical questions. In any case, we are reminded, technology will not create networks; only people can do that. So there is a chicken-and-egg problem: again, we did not emerge with a clear answer.

Whether in the schools or in the neighbourhoods, we need to recognise that society has an increasing choice of technologies. During the 1990s, we are promised, the personal computer linked into local and global networks, the cable, the video will be integrated in multi-media technologies. Which of the available technologies will prove most user-friendly to information-deprived groups? Our participants reported surprisingly encouraging results with some media, such as video. But the assessment may differ, depending on one's sphere of interest: the technology that proves best for the neighbourhood may not be so useful in resolving macro-problems, for instance. In dealing with very complex problems involving large-scale interactions, of which most environmental issues represent a classic case, a very large knowledge base usually needs to be provided and systematically analysed; there is thus a danger that the issues become simplified and vulgarised, using that last word in its pejorative sense. The opposite danger is equally real, though: it is that by providing progressively richer information, we may increase the efficiency of our decision-making processes, but only at the expense of equity. The division between information-rich and information-poor may become wider than ever before. Yet society needs to protect people's right not to use information and information technologies, if they so wish.

Finally, we returned to the question of values, which underlay so much of our discussion. One participant argued, with good reason, that this was a male-dominated conference which had ignored and excluded a whole range of sensual, social and cultural experience. Was this technology-driven world, she asked, one in which she would like her children to grow up? That squarely raised an old question in a new context: *quis custodiet ipsos custodes?* Who controls the controllers of the new technology? Is its introduction to be subject primarily to the market and to competition? Or could – and should – society determine its values outside the commercial nexus? Or is some combination of these two the possible and desirable solution?

We could all agree on the slogan that cities must be for people, and that technology must be people's servant and not its master. But, for these complex and highly politically-charged issues of control, a three-hour workshop was not enough.

Concluding issues

It is clear that the *technical* issues of access to information will not prove the major constraint in the transition to the informational society. Indeed, exponentially-accelerating technical development is likely to drive the entire process. In particular, the arrival during the 1990s of two fundamental information technologies, multi-media (hypermedia) and commercially-available expert systems, together with the increasing availability of broadband communication channels, will extend the technical potential almost beyond recognition.

The problems are not technical, but social and political. Whatever is done to extend access to information at every level – in the schools, at work, in the home, in mediatheques, in neighbourhood "one-stop shops" – there will continue to be huge differences in access to information by age, socio-economic group and cultural group, perhaps even (though this should diminish) by gender. This seems to demand the development of conscious policies to correct these deficiencies. The right to information, in other words, should become as basic as the right to equal access before the law, the right to pure water or the right to education, which most OECD countries achieved a century or more ago.

Related to this is the political point regarding the control of access to information. Here, there is less likelihood of unanimity. Though all OECD Member states are likely to continue with a mixture of public and private provision – sometimes, as in the case of Minitel or commercially-based television, in intimate relationship – there will undoubtedly be large differences in the precise mixture from country to country: some preferring to provide a whole range of services through monopolistic state agencies, some to throw them open to market competition. This is inevitable and even, in the interest of experimentation, desirable. But there are major issues concerning the effectiveness of the different formulae in preserving the right of access by the more deprived groups of society, which will repay further study by OECD.

REFERENCE

CASTELLS, M. (1989), *The Informational City: Information Technology, Economic Restructuring and the Urban-Regional Process,* Basil Blackwell, Oxford.

NEW TECHNOLOGIES
AND MANAGEMENT OF THE PUBLIC SECTOR

Costis TOREGAS
President of Public Technology Inc. (PTI)
UNITED STATES

Public enterprise as the management framework

The first international definition of the public enterprise framework was introduced during an OECD-sponsored conference in Copenhagen, Denmark, on Co-ordinated Information Systems (CIS), on 11-13 October 1989: "Implementation of CIS needs entrepreneurial spirit, be it provided by public, private, semi-private institutions or combinations of those entities. The concept of profit and rewards, which provides a general incentive to go beyond 'business as usual' in the private sector, may be considered an appropriate instrument. Revenues acquired through this instrument would assist in generating better services in the future through additional investment or in reducing the general tax burden". Indeed, the entire framework of public enterprise consists of two fundamental driving forces: the citizen as customer, and profits in the public sector.

A few weeks later, those concepts were sharpened during the First Annual Conference of Subsidiary Companies of Local Authority Associations, held in November 1989 in Herzelia, Israel, under the auspices of the International Union of Local Authorities, the Israeli Union of Local Authorities, Local Government Economic Services Ltd., and Public Technology Inc. As a consequence, Workshop B, focusing on a new framework for "public enterprise", was more than an isolated event; it concluded an important debate that had been underway for a considerable time, and that had mature and realistic conclusions to offer for implementation.

What exactly is this evolving, exciting and unconventional framework called public enterprise? Nothing more nor less than taking the notion of Orgware and expanding it to fit the realistic domain of the business world. Any technology, I believe, has three fundamental cost components (and commensurately it has three value components): hardware, software and orgware. Over 20 years of experience in the international effort to develop and implement local government technology have shown that hardware consumes roughly 5 per cent of the resources, software 15 per cent and orgware a perhaps surprising 80 per cent. It is this dominant component of local government technology investment that can form the foundation of the public enterprise model. Two themes animate the model: the citizen as customer, and an acceptable profit motive in the public sector.

These ideas are simple: as society becomes more and more oriented towards the individual under a series of increasingly decentralised options (many driven by technology advances), it is wise to follow the lead of private enterprise, for it has made a lasting shift towards a customer orientation. The customer of government is the citizen, so technology must first and foremost be viewed as a vehicle to serve that citizen better. Secondly, the wealth of technology that we find today can give rise to opportunities that can satisfy this customer orientation while producing opportunities for revenue at the local level. Selling geographic information in digitised form to transportation companies, providing municipal services through telephone networks and using software programmes that provide alternative financial payment methods such as credit cards are examples that come quickly to mind.

If such a discipline, combining a public sector orientation (towards the citizen) and a private sector orientation (towards profits) is to be effectively implemented and managed, it is necessary to create a new framework. And this framework is based on four Guiding Principles that can be seen as helpful in varied governmental structures and equally effective in different national contexts.

Competitive advantage

In any enterprise, whether public, private or hybrid, we should not be involved in areas where we do not enjoy a strong, unique and competitive advantage. Here are some examples of such competitive advantage for public entities.

- We make laws; if something doesn't work, or if an outmoded law or administrative mandate hinders economic activity, we can change it! The role for our elected officials is clear: to monitor the ever-changing environment and create a legislative framework that properly reflects a strong response. In the new world of public enterprise, laws themselves can be seen as an instrument of change and an opportunity for community activity.
- We not only command the production of the most complete information at the urban level, but we also have a very cost-effective system of maintaining those data on a daily basis, as the changes themselves either emanate from governmental action or are subject to governmental approval.
- We are strong if we stand united with one common, public sector-driven agenda for our societal concerns; the concept of market aggregation is an essential strategy with which to attack the fragmentation and expense associated with urban markets. In the United States, this aggregation role for the urban sector is being fulfilled by PTI, and the results are encouraging: several public enterprise efforts have already begun to produce profit streams back to the public entities under an aggregation banner (for one example of this principle at work, see Nations Cities Weekly: "Cities and Counties Profit with Pay Phones", 24 June 1990).

It is essential that urban leaders and the institutions that are responding to urban concerns on their behalf understand this principle of competitive advantage and exert a top to bottom review of all their traditional functions against those guidelines. In such a critical review, areas of strong advantage would become likely candidates for creation of a public enterprise activity.

Partnerships and new skills for partnership management

It is difficult to manage the exploding technological environment in cities and to continue providing quality services if we attempt to preserve all functions within the limits of government alone. Far easier and more productive is a model of co-operation and partnership with other sectors of the economy, as well as international elements.

But, in order for such partnerships to become effective, new skills of partnership management are necessary. Contract management around performance targets, evaluation and acceptance of risk factors and acceptance of other voices in the decision-making process of government are examples of new attitudes and skills that are essential tools of public enterprise.

In the information technology arena, partnerships are already playing a vital role: the development of new infrastructures such as fibre networks and software-defined networks offers excellent opportunities for public and private sectors to combine resources, reduce the overall cost and emphasize and use each other's strong points in successfully completing projects. These partnerships might take the form of joint ventures, franchises, contracts for services and many other alternatives depending on the regulatory and legal framework of the country involved.

Finally, in the context of international partnerships for effective technology management, we must not forget the importance of simple skills such as learning several languages for ease of communications, and establishing a network of contacts that can provide quick and accurate feedback regarding experiences with a given technology in other settings.

Equity and ethics

Equity stands as a constant reminder of the fairness with which we must approach the implementation of information technologies in cities. It is so easy to accept technological initiatives and forget the real human suffering underneath them. It is easy to ignore the deep chasm that has emerged between the haves and have-nots in this technologically rich society we have created in many places around the world. But we must not ignore it.

In fact, the management of technology under public enterprise can give us excellent tools – use of multi media technology, real time translation and other empowering technologies – with which to reverse this dehumanising and polarising trend. And, by wisely investing the profits that may accrue from a public enterprise approach to technology, we may even find a way to invest in the future of our communities rather than turn away from them and ignore those poor neighborhoods that may be locked out of the information revolution, and therefore be in greatest need of governmental assistance.

Unthinking the organisation

This is the most essential management lesson, yet the most difficult to accept: most governments insist on using technology to automate existing functions, thereby capturing forever on silicon chip and software the inefficiencies of the past and the shallowness of service delivery before the prowess of information technology. Indeed, the advantage of technology is that it can recast and dramatically change governmental service delivery along bold, new lines. Instead of considering individual, unco-ordinated services that are automated along agency lines, one has the ability to provide integrated service offerings along customer lines! And in doing so, the potential is there to satisfy the customer, decrease costs and perhaps find revenues hiding behind the public/private partnerships that can be used to animate such new instruments for the provision of services.

So what is there to prevent this vision from taking hold? Our own fear of the unknown, our own unwillingness to try something that is new! And it is imperative that we accept this challenge to unstick the past, unthink the organisation and think along customer lines, then bring in the technology that can best satisfy this customer orientation.

Discussion and recommendations

Against this background of public enterprise, an animated discussion was held; it covered management in general, as well as policy issues of information technology management, and served as an excellent stimulus for the workshop's subsequent recommendations. The most essential argument put forth from the audience was one of concern about policy leadership regarding technology.

Cities today are baset by many problems; we must find new patterns of evolution and revolution (one of the participants suggested that each city should have an officer in charge of revolution, *i.e.* the process of change we must all accept) that will permit the creation of new directions; and this is made more difficult in the current risk-averse atmosphere that pervades the city environment. The need for collective action will surely be underlined by the final conclusions; but, each city may also be seen as being on a new voyage to tomorrow... Information and Communications Technology (ICT) is like the fuel that will power up the starship City as she takes off to the stars! A better connected environment will nurture and help the citizen, a bold and decisive infrastructure will create new economic development opportunities for business and local government alike, a new way to address social ills. Yet, we have to be courageous and see who is in charge of this starship. No one? Faceless bureaucrate who are never in touch with the people? Or indeed an automated guidance system that hums inside a cold, gleaming computer shell?

Workshop B concluded that the most essential aspect of public sector management is to answer the question "who's in charge?" The word "vision" leaps to mind – the necessity for the elected officials of our cities to create a vision for the city that can then be empowered by ITC (and not the other way around). In order to implement the vision of our elected leaders, new roles and new structures will be necessary. Recognising the importance of the word "new", we recognised the significance of change management and agreed to make training a high priority: training of the citizens, training of the administrators, training of the elected officials in the new realities of the city.

Recommendations

We must look to individual city experiences and synthesize from those new roles and structures for cities; a Guidebook would help to bring together the diverse experiences under a common framework, and a more formal way to relate the city leaders and their staffs should be explored.

Roles for the new city include:

– partnership management of the extended enterprise that is today defining the new city (private industry, central government, academic institutions);
– joint investment in new technology development and the chance to receive a profitable return on that investment;

- defining priority needs that respond to real social issues;
- identifying the inherent competitive advantage of cities in new enterprises;
- opening up the city by providing its citizens with ease of access;
- appointing a manager of change (in physical, informational and human networks);
- appointing a training agent for its employees, leaders and citizens; (the new literacy elements are not alpha, beta, gamma but ISDN...);
- cross boundary roles in policy, management and financing.

Structures for the new city include:
- new, bold structures that enable cities to learn and then act;
- flatter organisational structures that reach significantly more contributors to the city's course (managing the extended enterprise);
- business relationships with private industry that form true alliances for progress (contracts, partnerships, joint ventures and beyond);
- new physical structures that merge bricks and electrons into powerful new edifices for progress (technoports, *etc.*);
- new spatial distribution patterns that satisfy people's demands and use ITC to amplify and facilitate service delivery where the citizens are.

We defined the city not only as a place of economic activity (under that definition, ITC can indeed be seen as destroying the city) but as a place where people converge to enjoy the good life together. In order to achieve this, we must be prepared to reorganise the old structures and build new ones that are motivated around the ''citizen as customer'' attitude. We must also remember the underprivileged in our cities. The role of ITC as generator of new wealth inside cities in some countries was recognised (in the public enterprise model of the United States, the ''citizen as customer'' and profits for cities take on added pertinence as ways to produce investment capital for our cities).

Finally, we saw four simple elements – social, cultural, technical and political – to guide us in this change. We are convinced that the leadership of cities will boldly lead their citizens in new and exciting dimensions, in partnership with central governments, private industry, academic institutions... and that the guiding light in this movement must be the strength of the individual, the human spirit that must always remain dominant over the technology.

NEW TECHNOLOGIES AND LOCAL ECONOMIC DEVELOPMENT

Rémy PRUD'HOMME
Professor at the Institut d'Urbanisme
de l'Université de Paris-Val de Marne
FRANCE

The city, the cradle of civilisation, has always been the place where information is produced, stored, used, traded and sold. A city is a commercial, financial and cultural centre where news on markets, financial data or ideas about the world in general are produced and transmitted. Venice, perhaps the archetype of the modern city, owed its prosperity to both the quantity and the quality of the information circulating within it – the availability of spices in Alexandria, the demand for wine in the United Kingdom, the political situation in Asia Minor. The heart of the city was not the port, where goods changed hands – it was the Rialto, where news was traded. A city is first and foremost an exchange. And exchanges have always been located in urban centres, since by definition their purpose is to gather together in the same place men and women with something to trade.

But is the supremacy of the city, or its monopoly, now under threat? If so, it is certainly not because information has become less important. Demand for information – or the amount of information produced – has never been as big as it is today. The share of services – that is, of information – in both final household consumption and intermediate consumption by firms is steadily increasing, at the expense of goods. It has been calculated that over 60 per cent of the labour force of the developed countries are engaged in jobs that consist primarily of handling information.

Yet the supremacy of cities may be jeopardised by new developments in the field of information technology. Clearly, information technology, like any other technology, has always had to adapt itself to innovation. The invention of the printing press and the telephone, for example, led to major changes in the way in which information is conveyed. The difference now is that the pace of change seems to have increased dramatically over the last ten years or so. Computers a hundred times more powerful than those available ten years ago can now be bought for a hundredth of the cost – computers with more capacity than those used by the Pentagon during the Vietnam war can now be found on supermarket shelves. Meanwhile, the transmission capacity of telephone networks has similarly increased by several orders of magnitude, as a result of the development, for example, of satellite systems and fibre optic cables. It is basically the marriage of telephone technology and computers ("enhanced telephone systems" and "supercomputers") that has given rise to the "new information technologies". Virtually unlimited amounts of information can now be transmitted to any destination at any time of the day, and at little cost. These new technologies have broken down the spatial and temporal constraints on information. The question that must now be asked is what that means for the city.

An attempt at an answer requires an economic approach, to ask what impact these new information technologies will have on the economy. The answer to that question is that they contribute to global economic development in two specific ways. First, they allow men and machines to produce more and better goods and, by eliminating waste, mistakes and repetitive tasks, they boost productivity. This is already the case in the automobile industry; tomorrow it will be a reality in health care and education.

Second, and more important, the new information technologies make it possible, or will make it possible, to reap the full benefit of two developments already in progress: the globalisation and liberalisation of world markets. National economies are fast becoming part of a global economy: trade is intensifying and most goods can now be bought and sold virtually anywhere in the world – with a few exceptions that the OECD is attempting either to eliminate or to limit. At the same time, and partly in conjunction with this trend, economies are becoming increasingly open, as the scope of regulations, licences, constraints and restrictions that governments impose is gradually shrinking. These two major trends have opened up dizzying vistas for economic agents in the options now open to them. But making those choices entails ''transaction costs''. This is where the new information technologies come into their own. They can help to reduce such costs, thus allowing economic agents to grasp the opportunities now on offer. Efficiency is based on freedom of choice. And exercising freedom of choice requires the use of information technology.

Yet the increased economic activity that will be generated by the use of new information technologies will not be evenly distributed. The increment will not be shared out pro rata, following the existing status of different sectors or different areas. At least three mechanisms are involved, all of which have implications for the location of activity – that is, for the future of cities. The new information technologies will give rise to new products, new processes and modes of production.

First, new products. The goods and services that will be produced and consumed will bear little resemblance to the present mix. Lower costs, improved quality and the appearance of new products (which cannot even be imagined at present) will revolutionise the structure of goods – and, above all, services – produced, and by the same token the spatial distribution of production facilities.

Second, new processes of production. The new information technologies will have a similar impact on production processes. Their effect on the production of goods is beginning to be seen: factory automation, increased specialisation of firms (resulting in the rapid development of freight transport), the replacement of raw materials and energy by information, and so on. The impact on services is less noticeable at present, and in particular on the two most important: health care and education. But there can be no doubt of the impact that information technologies will have on the geographical location of both people and activities.

Finally, new modes of production. New concepts, such as firm networks, or the standardisation of trade in information (the result of the development of electronic data interchange) are starting to emerge and to be used on a wider scale – again, with spatial implications.

Are these changes an opportunity for cities (and if so, for which ones?), or are they, on the contrary, a threat to their continued existence? Will these new technologies attract people to cities, or will they drive them away? Do they both intrinsically and automatically strengthen the pull of cities, and in particular large cities, or do they encourage activities to spread into smaller urban centres, run-down suburbs or rural areas? At present, no one can confidently predict the outcome.

It can be argued that the increased importance of information together with easier flows of information (the two major advances offered by the new technologies) will allow both firms and people to operate from any location, and will thus lead to the decentralisation of activities and a concomitant decline in the importance of cities or, at any rate, the major cities. Teleworking – working from a computer terminal in the home linked to a large number of existing data banks, in a calm and pleasant location, far from the noise-ridden, polluted environment of the city – will become easier and far more popular, emptying the cities of a least some of those involved in handling information. Teleconferences and fax machines will do away with meetings and the concentration (on either a permanent or temporary basis) of economic agents in the same, and obviously urban, location. In short, there will no longer be any call for exchanges located at the centre of the city, since their work can now be carried out by a host computer located anywhere.

Conversely, it can be argued that the growing volume of activities in creation, design, training, organisation, management, co-ordination, information, and also recreational, cultural, symbolic, social and religious activities, will generate an increasing demand for people to speak, meet and trade with one another, a demand which can be satisfied only by even larger cities. A business merger cannot be negotiated over a computer link any more than psychoanalysts can conduct their consultations over the telephone. As one of the participants put it, "It's hard enough as it is to manage the people you can see; just imagine trying to manage people you can't see!" In this light, cities would seem to be more necessary than ever, both as the location for productive activities and as the location for household consumption.

Developments over the last few years would seem to confirm the second approach – as does the somewhat older example of the telephone. Far from reducing interpersonal contacts, the telephone if anything would seem to have increased them. And teleworking, which the futurologists of the 1970s predicted would become a major growth area, has so far remained marginal.

At the same time, a counter-trend has been witnessed in which the relative attractiveness of major cities, and city centres, has grown, as shown by the increase in property prices. They reflect the interest shown by firms and households in locating in a given spot, and hence the attractiveness of that location. They thus have clearly always been higher in large cities than in small towns, and higher in the centre of these cities than in the suburbs. But the difference between property prices in large cities and those in small or medium-sized cities, that is, the attractiveness of large cities over small cities, has not always been constant. This difference, which shrank in the 1970s and early 1980s, has apparently grown steadily wider since 1985, suggesting that major cities, and particularly large conurbations like London, New York, Tokyo or Paris, are becoming increasingly sought after as locations.

Two conclusions emerge from predictions of future trends, although they have to be treated with caution. The first is that there is undoubtedly a wide range of possibilities. The new information technologies do not militate openly for or against cities. They simply erode geographical determinism. They are just as likely to encourage as discourage the concentration or dispersion of people and actvities. They simply offer people a choice, regardless of whether they live in a major conurbation or a rural area.

The second is that these technologies can be dangerous. A first reaction might be to think that some cities, notably large metropolitan cities, would be better placed than others to seize this opportunity, given their larger initial resources in terms of skilled workers, financial assets and, last but not least, information. But even though the precise impact of the new technologies cannot be determined, they would seem to attract more people towards cities than they encourage to leave. It is therefore possible, and indeed probable, that they will help to draw people into large cities. Fears of rapidly decaying cities would therefore seem to be unfounded. Yet any relief this prompts may prove to be short-lived, since renewed urban growth is accompanied by at least three serious threats to society.

The first danger is political. Large metropolitan cities are already the wealthiest areas in any given country. Renewed growth would merely exacerbate existing differences between richer and poorer areas of every OECD country, which would be hard to justify in political terms. Over the last 50 years, these differences, which are often referred to as "regional disparities", have slowly but surely narrowed. One of the main mechanisms behind this erosion has been the tendency of firms to locate in areas where salaries are lowest, in areas which, by definition, are the poorest. Firms will always want to locate in areas where salaries are lowest, but nowadays they are even keener to locate where information is plentiful or, to be precise, where there are people capable of exploiting information, that is, in large cities and the wealthiest regions. Because of the growing importance of information, the mechanism which automatically reduced regional disparities has been displaced by another mechanism which automatically exacerbates them.

The second danger is physical. The major cities are already beginning to choke to death under the combined impact of congestion and pollution. Unchecked growth would soon make them unmanageable, to the benefit of no one. Congestion and pollution are classic externalities. By locating in a city because that is where it finds what it requires, a firm (or a household) imposes costs on others which it does not have to pay itself and which are larger than the benefits it receives. The costs are borne by society.

The final danger is a social one. The exploitation of information and the new information technologies calls for training, effort and talent. It requires a certain expertise that not everybody is able to offer. The development of information technology, particularly in urban areas, when cities in the developed world must contend with rising immigration from poorer countries, may well widen the gaps that already exist between different social groups. "Two-speed" cities may emerge. The attack by two unemployed youths on a "yuppy" woman in Central Park in New York caused such an outcry because it was more than a news item: it had symbolic value.

This brief overview of the opportunities which the new information technologies offer cities, as well as the dangers they represent, naturally prompts a number of recommendations for economic policy. The analysis suggests that economic policy-makers can influence developments, maybe more in this area than in others, and that is exactly what they are doing. These recommendations are addressed to central governments, regional authorities, and also, perhaps, to firms.

Clearly, central governments must help the development of the new information technologies; to a certain extent this development will depend on their decisions and their investment. These technologies are generally profitable, so in most cases it is both unnecessary and undesirable for them to be funded by the public purse. But this is not always true and, indeed, in some respects these technologies are potentially monopolistic and should therefore be regulated. The problem is how to regulate the sector without stifling it.

Above all, central governments must stiffen their regional development policies. If current developments exacerbate regional disparities, as feared, they will have to be countered by vigorous policies designed to support regions which are either lagging behind or are under threat. These policies will be harder and more expensive to implement since, instead of supporting a mechanism, they will now have to reverse a trend – and it is always harder to sail against the wind than with it. Yet such policies are essential. If central government does not maintain a certain balance between the regions and between cities, who will?

Local authorities – regional bodies, cities, intermediary organisations – will have a major role to play in introducing and supervising the new information technologies. The answer to whether the state, local authorities or firms should take the lead would seem to be the local authorities. Almost all the initiatives described and discussed at the OECD conference were sponsored by local authorities. So what are they doing? And what can they do?

First, they can encourage firms producing or using new information technologies to locate in their area. But there is a paradox here. These technologies, which allow their users freedom of location, are starting to be adopted by firms which are themselves firmly rooted in a given location and which have a higher growth rate than other firms, either because they produce the technologies or because they use them. This fact has not escaped the notice of town planners who have accordingly set up teleports or specialised zones.

Local authorities can then use the new information technologies themselves to manage municipal public services. They are not doing this simply to set an example: although the introduction of new information technologies may well encourage unchecked, and ultimately unmanageable, urban growth, it simultaneously offers the means to manage ever-larger cities. New information technologies are both the sickness and the cure. There are many examples of this phenomenon. Road pricing, which makes motorists pay for the congestion they cause – the economist's dream – has now been made possible through the development of on-board electronic systems. Soon all drivers will be receiving traffic bills alongside their telephone bills, which will make sure that they do everything in their power to reduce the congestion for which they are responsible, and which will thereby help to improve traffic flows (providing that the price is high enough) more than any regulations that can be devised could ever do. Another example is in preventing damage to property or personal injury. Videomonitoring and the transmission of electrocardiogram data, coupled with the provision of emergency response teams, offers possibilities that are only just beginning to be exploited.

Cities must subsequently concentrate their efforts on training. Poor or non-existent training facilities are a major obstacle to the development of new information technologies, and are probably the principal cause of the emergence of ''two-speed'' cities. To attract firms and ensure social stability, it is better to have trained workers without a teleport than a teleport without trained workers.

Local authorities, as many delegates to the OECD conference stressed, must also beware of trends in fashion, which might prompt them to devote all their resources to the development of new technologies while abandoning or neglecting more conventional programmes or infrastructure. Social aid and public transport – whose efficiency can be improved by using these new technologies – must continue to receive their full attention and a large portion of their budget allocations.

Firms, particularly large firms, will undoubtedly be called upon to play the part incumbent on them in developing the city. Obviously, their main task is to lower costs, within the confines of the law, and to create jobs, and in this respect the imperative of turning a profit will force them to meet the challenge. But is this enough? Can one now accept the classical and strict division of labour between an economic power which produces goods and services and a political power which regulates that production? The interests of the public and private sectors coincide so closely in new information technologies that the question simply cannot be avoided. The decisions taken by firms have such an impact on the life of the city, and the decisions taken by politicians have such an impact on the profits made by firms, that the old philosophy, ''every man to his trade'', would probably be inadequate both for firms and for cities. Many industrialists would be the first to admit the truth of this statement, and to want to play a bigger role in the life of their city or their region. The forms and rules for such participation concern everyone.

C. Conclusions

SYNTHESIS AND RECOMMENDATIONS

Georges MERCADAL
President of URBA 2000
FRANCE

It was important for this seminar to be held, and for it to attract a relatively large audience: some 320 people have taken part. It was important for this event to be really international: only half the participants are French, and the remainder come from another 22 countries. And it was also important for local representatives and mayors to take part in this meeting, which they have done.

This fully representative symposium shows how the new technologies are sweeping like a wave through all aspects of urban life, from traffic to health, to places and methods of work, education, the environment, utilities, safety, neighbourhood life and the home itself. The bird's-eye view we have obtained of all these aspects in two and a half days has brought out the extent and speed of a development which escape our notice in the daily routine of our responsibilities or our particular job.

At the same time, the proceedings have raised a critical issue, for nobody can say that this new wave of technology will automatically bring greater well-being to cities. It could just as well lead to even greater fragmentation, to quote the Secretary-General of the OECD, or to even more divisions in society, to quote the words of the French Minister Emile Biasini or Yves Dauge, the French Government Delegate for Cities.

This meeting further shows the wave of new technologies sweeping through all the functions performed by cities, and which may well make all the problems of urban civilisation worse.

In response to this risk, the policy statements made at this meeting by Ministers and former Ministers, mayors and elected officials have clearly defined the course of action as one based on humanism and resolve.

Resolve, because as everybody has stressed, techniques are neither good nor bad in themselves, and only political resolve can tip the scales in favour of what is good.

Humanism because this resolve must be aimed at serving man. "Science without a conscience only destroys the soul", as Mr. Biasini reminded us.

Today, reinventing the city means serving mankind.

Reinventing the city means opposing the trends towards the fragmentation and disintegration of society which generate the two-tier society roundly condemned by all the speakers.

In addition to the conclusions which have been drawn concerning the economic, practical and social roles played by cities, the overview given by this conference enables us to define a general concept: with the advent of information techniques, cities have an historic opportunity of again interconnecting all urban functions by means of horizontal links capable of directly resisting the trends towards the disintegration of society.

This concept can be implemented only if action is taken at two levels: individually by each city, and at government level as well.

Detailed recommendations have been put forward for each of these levels.

Recommendations for cities

Cities cannot afford to look on passively while the wave of new technologies is breaking. They must be energetic, and simultaneously act as developers and catalysts while ensuring quality. For this purpose they must set up and manage five horizontal functions:

Access to telematic systems

Cities should provide or facilitate access to these systems, and also monitor user standards for the customer/citizen.

The systems are to be understood in the broad sense.

Their main component consists of voice, data and image telecommunication networks. So far the vast majority of cities have shown interest only in cable television. The various papers pointed out that they must also turn their attention, as Amsterdam and Osaka have done, to links with other cities around the world if they wish to have any chance at all of attracting efficient firms and telework.

A city's internal transmission networks which can be used for remote management of water, sewage, transport and other utilities and are also a means of creating systemic effects between these management operations, as suggested by Gabriel Dupuy, will fall into a different technical (and especially administrative) category from cable TV. Even today few mayors have tackled this subject, which will very probably lead them to negotiations with their country's PTT authorities.

The second component in the systems is the smart card. The paper on health brought out all its advantages. There can be no doubt that it will be used in the same way, particularly as a smart means of payment capable of taking various kinds of reduction into account, for transport and all other urban activities. Rather than have a countless number of cards, each under-used (and therefore mis-used), it seems infinitely preferable for a municipality itself to set up a multi-sectoral card distribution and reader network, especially since card storage capacity now makes it easy to create a multi-purpose card.

These urban systems also include the terminals which each sector will be using, and cities are therefore responsible for quality control. Workshop A and Peter Hall in his conclusions stressed how essential user-friendly terminals are if the new technologies are not to erect another barrier between the haves and the have-nots. Experiments to date have shown that the best intentions in the world could finally come to grief because of a lack of user-friendly terminals.

The municipal authorities must somehow or other ensure they have a say in the user-friendliness of the terminals provided in their cities, whether by transport operators, security companies, telework centres, *etc.*

The "co-ordinated information system"

A few years ago everybody was thinking in terms of a vast central file for all the information needed to run a town. This was known as an urban databank, or an integrated management system.

At the Copenhagen seminar we all acknowledged that this ambition was now obsolete because there are now actors in every city who all have their own files, their own software and their own computers. They cannot be asked to opt for a completely different solution. But as the Copenhagen seminar showed, it is technically feasible to interconnect them and let them communicate with one another at a reasonable cost. It is up to the mayor to insist that all the systems being set up to manage particular functions communicate with one another in the same language, with each being left in complete control of its internal configuration.

This second task is sub-divided into two. First, develop this common language as a set of standardized messages; then, give the actors added values enabling each sector to be enhanced by the information available in the others and achieve the systemic effects mentioned by a number of speakers, particularly Gabriel Dupuy.

"Front-of-house assistance"

This down-to-earth term came into being some time ago when airlines noticed that many of their customers were at a loss in airport concourses, not knowing which counter to go to. So "front-of-house" help was provided. It is similar to what William Hansell called "empowering" or "enabling" the citizen, and something endorsed by the Mayor of Athens. It necessarily comprises a number of aspects.

First, the consumerism aspect, meaning that the consumer has to be given sufficient information to stand up to the monopolies which claim to serve him. I run the Société des Eaux de Paris. By typing on the Minitel keyboard you can find out if the water in Paris contains nitrates or pesticides. The Minitel system receives its information from an approved laboratory that is completely independent of my company, which is a guarantee for the consumer and an incentive for the producer.

The same can be done in every field of urban life since the new technologies enable us, at reasonable cost, to measure quantities and qualities and make them known. This is a direct answer to the Mayor of Athens, who called on us to decentralise municipal control.

Second, the acquisition of knowledge. Gilbert Paquette provided some basic information on this subject. Personally I am quite sure that this is the major piece of news at this Seminar, enabling us to tell Yves Dauge that he is not alone in the run-down districts and that the new technologies have something to offer him.

I found two remarks by Gilbert Paquette of particular interest. The first was that information technologies made it possible to personalise teaching methods themselves and thus get through to those who no longer hear or understand. The second remark was that the combination of communication and information technologies enables us to get through to the individual in his normal environment, "front of house", and particularly at work, at home, or indeed at school.

Gilbert Paquette suggested that a Municipal Office should be set up to deal with this kind of training. I think that this suggestion should be considered seriously and taken further. This is really an "enabling" and "empowering" idea. Storage miniaturisation, communication capacity and artificial intelligence should enable us to set up aids for individually tailored acquisition and handling of knowledge.

Those who might have misgivings about the term "Office" should be reassured: the role of this body is not to replace school, vocational training or cultural facilities. It must be stressed that it is to provide "front-of-house" help with all these services; it is to help the individual to participate in the labour market, obtain access to the city's cultural facilities, and to enhance the teaching he receives at school (we are thinking, for example, of the experiments in helping foreign pupils in their homes).

Lastly, there is a question of integration in neighbourhood life. It is true that town planners have been dreaming about this for a long time: about getting back to what Peter Hall has just referred to as the urban village. Who would not like to live in Greenwich Village or Montmartre? Can the new technologies help us to transform this dream into reality? Perhaps. I must admit to no more than measured optimism here.

The facilitation of democratic processes

I shall not make any comments on this point. But I mention facilitation because Mr. Giraud referred to it a number of times yesterday. Interactivity, he said, is democracy. I shall leave it to you to judge. I sincerely hope it is. It is up to the elected officials to prove to us that this can actually become a reality and that the new communication techniques can help us to weather the crisis in representative democracy.

Adopt a consistent refocusing strategy

It is not enough to tell cities to be energetic and master the new fields of action that the new technologies are opening up. They are already saturated with responsibilities, they often employ a very large staff, and the tax burden they place on their citizens is often close to its upper limit.

To succeed, cities must adopt a refocusing strategy which will transform their style, organisation and behaviour.

This transformation calls for a threefold approach:

Act like a private holding company

Private holding companies have shown that one of the basic requirements for success is the constant refocusing of activities. There is nothing to be gained by dispersal of effort; it is necessary to concentrate on essentials.

Many activities which are now carried out directly by municipalities themselves could be very largely delegated and decentralised. This is the case of all mature sectors – water, sewerage, gas and electricity, telephones, traffic, transport and the removal of household refuse.

These are mature sectors because they can be directly monitored by the citizen as we suggested above, thereby enabling the elected representatives to play a more strategic role, and because practically all of them can be run on an economically and financially independent basis through systematic charges.

Any private holding company makes money from mature activities and invests it in sunrise activities. Cities should also make money, *i.e.* reduce the workforce employed in mature activities – water, sewerage, the road system, *etc.* – and invest in sunrise activities such as "co-ordinated information systems," 'front-of-house' assistance, "telesecurity, *etc.*". In the future these are the services which will mark the difference between a good mayor who gives his electors something extra and one who only sees to needs.

It is certain that such basic changes cannot be made without difficulties, and the main one certainly concerns municipal staff. In France, and probably in all OECD countries, their rules of service do not encourage mobility and change. But even in this area, the use of the new technologies provides some answers: reductions in the workforce are usually accompanied by an increase in qualifications and therefore in salaries, which may give enough incentive to put up with the changes.

Enlist the spirit of enterprise

Cities must also enlist the spirit of enterprise. This point was very well analysed at the Copenhagen seminar. How was a "co-ordinated information system" to be set up?

A number of strategies were considered, ranging from the one which originates with central government and lays down the law, to the one which consists in waiting for initiatives to emerge. One aspect was considered essential, whatever the strategy adopted, and this was the spirit of enterprise. And nowhere has this spirit of enterprise been better illustrated than by Geneva Canton's Land Register Department, a public body which is now managing to sell the information it creates.

This means that the spirit of enterprise is neither public nor private, that its legal trappings may vary greatly with the country but that it is absolutely necessary as an input.

Mark the new policy with visible symbols

A city's commitment to an active and consistent policy of using the new technologies must be expressed visually and physically. I am not really one of those who think that the city should not be seen. It must provide the citizen with visible symbols if it is to play its role fully as a cultural centre.

Places must be created which give everybody a sense of reality. As a Japanese friend said to me yesterday evening, "we are afraid in Japan that our children's only possible contacts with the world will now be through the computer screen and keyboard." Is the danger imminent everywhere or not? In any case, the risk is too great to run.

But these places must also offer the citizens symbols, values and emotions, which is the role of architecture. And if they do, it will be the best way for the municipal authorities to show their citizens that technology has not gone to their heads and that they know, as the Mayor of Athens said, that "although many things change, many others remain the same."

The Musée de la Villette where this conference is taking place is certainly one of these new places symbolic of tomorrow's city. Is it the ultimate in its architectural expression? The future will tell. But what is certain today is that such a place is a highly significant expression of tomorrow's city, for it is a place for people to meet, and that alone: as everybody who has had a chance to take a look at the surrounding area has been able to observe, a place of this kind has an irresistible power to restructure its surroundings and leaves its mark on the entire town.

Increasingly, in a trend that the new communication technologies will speed up, the city is a place where people want, rather than have, to go. In addition to creating places, buildings and architecture, the city authorities will be organising events capable of attracting a wide public.

We already have examples, ranging from the "pop" music events which were all the rage about ten years ago to the music festivals or major celebrations which are now an accepted part of city life. The city is a stage that artistes use to put on for some time a show that attracts hundreds of thousands and even millions of spectators, and this can again be done with the use of information technologies.

As Peter Hall said in his conclusion that "schools must change their nature and chiefly become a place where children meet and talk with one another, meaning that what was until recently a secondary aspect will become the main feature".

It is just the same with cities. And being together constantly is not always possible. Togetherness is now an extremely basic event and the new technologies can help us to create, and especially manage, such events.

Recommendations to governments

One striking point about this seminar is that nobody has called on central government to promote the new technologies for cities. Municipal responsibility for tackling the problems stated at the outset was accepted and unquestioned. Another possible explanation is that everybody in the western world has become accustomed to government's deliberate policy, which is a reaction to the period running from the post-war years to the start of the 1970s, of leaving cities to their own devices; the United States and the UK are only extreme examples of this policy.

All the same, a number of European countries in recent years have appointed Ministers or high-ranking officials with responsibility for cities. The inner cities have been acknowledged as a national issue in the United Kingdom. The idea is again emerging, and will certainly make headway, that the city is a mechanism which has always integrated the citizens in economic and social life. If this mechanism stops working, if it no longer serves its purpose or costs too much, then the nation as a whole suffers.

It thus seems to be in keeping with the emerging trend to urge governments to create the general environment and conditions over which cities have no control but which they need in order to succeed in actively using the new technologies.

We shall make three recommendations:

Promote research activity

Research should deal first and foremost with user-friendly terminals, for as Workshop A stressed this is an essential requirement to ensure that young and old, poor and rich, educated and uneducated, all have access to information networks.

But research must also be directed at reliable and inexpensive measurement instruments for all urban services. These are needed if the decentralised control which has been demanded is to succeed.

And although the question was not really discussed at the conference, it must not be forgotten that intensive use of the new technologies is based on attitudes that do not come the most naturally to man: for example, the objective of rationality and planning of behaviour, or the possibility of getting used to artifacts and to indirect contact with reality.

Conversely, if man can adjust to and develop these attitudes in order to use screens and keyboards, will something somewhere in him not be destroyed? What kinds of compensatory behaviour might then be triggered? Is there no danger of these instruments being used for more effective control of a population that has become more passive?

The social sciences certainly have their work cut out if they are to throw light on these questions, and this will be the third line of research to be developed.

Encourage centrifugal movements

No social problems will be resolved in the outlying districts of the major capitals or in the outlying cities of countries and continents unless jobs are provided.

In the OECD countries, the jobs will be increasingly in the tertiary sector.

Thus the question discussed by Workshop C, whether the new technologies would encourage decentralisation or reconcentration, is a basic one.

The answer that a laisser-faire policy will result in a new surge of concentration is a serious threat that all the imbalances now affecting cities will become greater. It must be seen by governments as a real danger signal.

The experts' opinion – that the die is not yet cast, and that a firm and steady resolve can take the economy in the opposite direction, since it will benefit fully from the freedom in business activity location given by the new technologies and advances in transport – must therefore be seen as an urgent recommendation.

The idea has been expressed that government should finance pilot studies on alternative locations for a number of public and private administrative bodies. It has also been said that in Europe these studies should be conducted at European level, particularly for the benefit of the cities which in each country act as regional capitals in relation to the main economic hub (Birmingham, Amsterdam, Munich, Lyons, *etc.*).

Facilitate co-operation between cities across the world

If, as a number of speakers insisted, we are to control and not be controlled by the new technologies, municipalities have to co-operate at world level and harmonize their viewpoints. For the productive system in this area consists not of small local enterprises but of international groups. Thus if demand remains dispersed, the logic of technology and industry will predominate to the detriment of the citizen's real needs.

Three tasks emerge which cannot be properly carried out by a single city or even all the cities in a single country, and must therefore be tackled at international level:

- full and thorough exchanges of information on the experiments carried out by various cities and the evaluation of their results, particularly of the benefit to the user;
- functional standardization of the products and services expected of the new technologies, in co-operation with industry if necessary;
- evaluation of the large-scale systems marketed by major groups with the objective of worldwide distribution.

Cities do not have an adequate international system to organise these tasks.

A number of conference participants have suggested that OECD take the initiative of setting up a city club that could be the start to a permanent consultation structure whose final form has still to be defined.

THE ROLE OF CITIES IN SUSTAINABLE DEVELOPMENT: A CALL FOR NEW PARTNERSHIPS[1]

Siegfried BRENKE
Head of Urban Affairs Division, OECD

I will begin my concluding remarks by responding to the questions raised by Mr. Mercadal in his final presentation, and specifically in his last recommendation to governments, urging them to "facilitate co-operation between cities across the world".

Yes, I certainly agree that we need to *standardize,* simply for the sake of achieving economies of large scale production and for international networking.

Yes, we need to *exchange* information about available technological options better so as to fully exploit the potential for improvements.

And finally, yes, we need to *evaluate* systematically and comprehensively these technological options to make sure that the process of technological exchange is also a *progress* in achieving our objectives.

The challenges the world is facing today are without precedent, both:

– in safeguarding the environment, not only in OECD Member countries but worldwide; and
– in overcoming poverty, not only in the developing world, but also in OECD Member countries.

To meet these challenges we not only need the best available technologies. We also need *new partnerships* between:

– East and West;
– OECD Member countries and the developing world;
– the public and the private sector; and
– different government agencies and different tiers of government.

It is in this area that the OECD's Urban Programme can make a contribution and perhaps even provide a unique opportunity.

With the approach of the *"Global Village"* urban affairs are no longer "internal affairs". More international co-operation is needed not only to agree on standards and structures of communication. We need the Urban Observatory to compare and evaluate success and failure and we need a functioning network to exchange this information quickly. Only then can intelligent solutions be turned into broad-based innovation and worldwide economic restructuring as quickly as it is needed.

Change and innovation *involve risk* but so do avoiding change and postponing innovation. Risk is an unavoidable part of economic and political decision-making and there seems to be little progress, if any, without taking a risk. However, we try to reduce risk where possible. National governments tend to be especially "risk averse" and probably they need to be.

Changing policy directions, restructuring the economy by fully integrating environmental costs in pricing mechanisms and designing new systems of empowerment and social security on a national level is like changing the course of an oceanliner in unsafe waters: it is dangerous, it takes time and once a decision has been taken it is equally difficult to reverse the new course. It is much easier, faster and less risky to man25vre a smaller boat. Why can't we look to cities to play the role, in the national and international context, of these escort ships testing the waters for the mothership and the international fleet of oceanliners?

Cities develop their unique approaches by responding to their specific needs and exploiting their specific capacities, and cities are voluntarily experimenting with new solutions. Successful *cities* may be *leading* the way – not in theory but *by example*. Clearly, in finding out what works and what does not in practical application, national governments need local leaders as partners to further international co-operation.

Again, OECD's Urban Programme may be able to offer its collaboration. Its framework for international co-operation exists on the national level and the Programme also offers opportunities for working more closely with subnational governments and directly with municipalities.

The new Mandate for the Group on Urban Affairs: *"The Role of Cities in Sustainable Development"* provides a framework for co-operation which puts local actions into the global context. It defines the success of each city by its contribution to a long-term economically sound, socially harmonious and environmentally feasible development of the Global Village.

This may be seen as an utopian goal and possibly it is, but which one of our basic principles is not. Mr. Tritsis, Mayor of Athens, has drawn our attention to the Greek origin of the word "utopia" meaning "nowhere" or "no place for the idea", and has defined his task as making a trip from "utopia" to "entopia", meaning a *place,* a locality, in which the idea can become real.

Thus, I would like my closing thoughts to be seen as an encouragement to all of us to make a further step forward *from "utopia" to "entopia"*.
- Let us *work together* to define more precisely the role of cities in sustainable development and let us do so in the international context and in close co-operation with cities;
- Let us *turn definitions into advice* and help to develop guidelines based on international experience and local practice;
- Let us *turn advice into broad-based action* through intensified international co-operation, extended municipal linking and close co-operation between different agencies and tiers of government; and finally;
- Let us *speed up and improve action by* making the fullest *use* possible *of the technological support* available, without which the challenges we face cannot be met in time.

NOTE

1. The opinions expressed in this paper are those of the author and do not necessarily represent the views of the OECD.

AFTERWORD

The impact of the development of information technologies and of telecommunication on the functioning of cities was part of the OECD Group on Urban Affairs' programme of work in 1987-90. The work has been specifically marked by two international conferences which were held in Copenhagen (Denmark) in October 1989 and in Paris (France) in November 1990.

The Conference in Paris had a general orientation. That is why this conference presented a broad range of themes and examples. Organised in co-operation with the Délégation française à la Ville and URBA 2000, it provided the occasion for bringing together a large number of Mayors and local managers, scientists and participants from industry and business of OECD Member countries. These presentations are the subject of the present volume.

The Conference in Copenhagen, organised by the Government of Denmark, focused on the development of co-ordinated information systems and their application to improve urban management and the quality of services provided to different groups of the urban society. That theme will be taken up in another volume of this OECD study.

WHERE TO OBTAIN OECD PUBLICATIONS – OÙ OBTENIR LES PUBLICATIONS DE L'OCDE

Argentina – Argentine
CARLOS HIRSCH S.R.L.
Galería Güemes, Florida 165, 4° Piso
1333 Buenos Aires   Tel. 30.7122, 331.1787 y 331.2391
Telegram: Hirsch-Baires
Telex: 21112 UAPE-AR. Ref. s/2901
Telefax:(1)331-1787

Australia – Australie
D.A. Book (Aust.) Pty. Ltd.
648 Whitehorse Road, P.O.B 163
Mitcham, Victoria 3132   Tel. (03)873.4411
Telefax: (03)873.5679

Austria – Autriche
OECD Publications and Information Centre
Schedestrasse 7
D-W 5300 Bonn 1 (Germany)   Tel. (49.228)21.60.45
Telefax: (49.228)26.11.04
Gerold & Co.
Graben 31
Wien I   Tel. (0222)533.50.14

Belgium – Belgique
Jean De Lannoy
Avenue du Roi 202
B-1060 Bruxelles   Tel. (02)538.51.69/538.08.41
Telex: 63220   Telefax: (02) 538.08.41

Canada
Renouf Publishing Company Ltd.
1294 Algoma Road
Ottawa, ON K1B 3W8   Tel. (613)741.4333
Telex: 053-4783   Telefax: (613)741.5439
Stores:
61 Sparks Street
Ottawa, ON K1P 5R1   Tel. (613)238.8985
211 Yonge Street
Toronto, ON M5B 1M4   Tel. (416)363.3171
Federal Publications
165 University Avenue
Toronto, ON M5H 3B8   Tel. (416)581.1552
Telefax: (416)581.1743
Les Publications Fédérales
1185 rue de l'Université
Montréal, PQ H3B 3A7   Tel.(514)954-1633
Les Éditions La Liberté Inc.
3020 Chemin Sainte-Foy
Sainte-Foy, PQ G1X 3V6   Tel. (418)658.3763
Telefax: (418)658.3763

Denmark – Danemark
Munksgaard Export and Subscription Service
35, Nørre Søgade, P.O. Box 2148
DK-1016 København K   Tel. (45 33)12.85.70
Telex: 19431 MUNKS DK   Telefax: (45 33)12.93.87

Finland – Finlande
Akateeminen Kirjakauppa
Keskuskatu 1, P.O. Box 128
00100 Helsinki   Tel. (358 0)12141
Telex: 125080   Telefax: (358 0)121.4441

France
OECD/OCDE
Mail Orders/Commandes par correspondance:
2, rue André-Pascal
75775 Paris Cédex 16   Tel. (33-1)45.24.82.00
Bookshop/Librairie:
33, rue Octave-Feuillet
75016 Paris   Tel. (33-1)45.24.81.67
                (33-1)45.24.81.81
Telex: 620 160 OCDE
Telefax: (33-1)45.24.85.00 (33-1)45.24.81.76
Librairie de l'Université
12a, rue Nazareth
13100 Aix-en-Provence   Tel. 42.26.18.08
Telefax : 42.26.63.26

Germany – Allemagne
OECD Publications and Information Centre
Schedestrasse 7
D-W 5300 Bonn 1   Tel. (0228)21.60.45
Telefax: (0228)26.11.04

Greece – Grèce
Librairie Kauffmann
28 rue du Stade
105 64 Athens   Tel. 322.21.60
Telex: 218187 LIKA Gr

Hong Kong
Swindon Book Co. Ltd.
13 - 15 Lock Road
Kowloon, Hong Kong   Tel. 366.80.31
Telex: 50 441 SWIN HX   Telefax: 739.49.75

Iceland – Islande
Mál Mog Menning
Laugavegi 18, Pósthólf 392
121 Reykjavik   Tel. 15199/24240

India – Inde
Oxford Book and Stationery Co.
Scindia House
New Delhi 110001   Tel. 331.5896/5308
Telex: 31 61990 AM IN
Telefax: (11)332.5993
17 Park Street
Calcutta 700016   Tel. 240832

Indonesia – Indonésie
Pdii-Lipi
P.O. Box 269/JKSMG/88
Jakarta 12790   Tel. 583467
Telex: 62 875

Ireland – Irlande
TDC Publishers – Library Suppliers
12 North Frederick Street
Dublin 1   Tel. 744835/749677
Telex: 33530 TDCP EI   Telefax: 748416

Italy – Italie
Libreria Commissionaria Sansoni
Via Benedetto Fortini, 120/10
Casella Post. 552
50125 Firenze   Tel. (055)64.54.15
Telex: 570466   Telefax: (055)64.12.57
Via Bartolini 29
20155 Milano   Tel. 36.50.83
La diffusione delle pubblicazioni OCSE viene assicurata
dalle principali librerie ed anche da:
Editrice e Libreria Herder
Piazza Montecitorio 120
00186 Roma   Tel. 679.46.28
Telex: NATEL I 621427
Libreria Hoepli
Via Hoepli 5
20121 Milano   Tel. 86.54.46
Telex: 31.33.95   Telefax: (02)805.28.86
Libreria Scientifica
Dott. Lucio de Biasio 'Aeiou'
Via Meravigli 16
20123 Milano   Tel. 805.68.98
Telefax: 800175

Japan – Japon
OECD Publications and Information Centre
Landic Akasaka Building
2-3-4 Akasaka, Minato-ku
Tokyo 107   Tel. (81.3)3586.2016
Telefax: (81.3)3584.7929

Korea – Corée
Kyobo Book Centre Co. Ltd.
P.O. Box 1658, Kwang Hwa Moon
Seoul   Tel. (REP)730.78.91
Telefax: 735.0030

Malaysia/Singapore – Malaisie/Singapour
Co-operative Bookshop Ltd.
University of Malaya
P.O. Box 1127, Jalan Pantai Baru
59700 Kuala Lumpur
Malaysia   Tel. 756.5000/756.5425
Telefax: 757.3661
Information Publications Pte. Ltd.
Pei-Fu Industrial Building
24 New Industrial Road No. 02-06
Singapore 1953   Tel. 283.1786/283.1798
Telefax: 284.8875

Netherlands – Pays-Bas
SDU Uitgeverij
Christoffel Plantijnstraat 2
Postbus 20014
2500 EA's-Gravenhage   Tel. (070 3)78.99.11
Voor bestellingen: Tel. (070 3)78.98.80
Telex: 32486 stdru   Telefax: (070 3)47.63.51

New Zealand – Nouvelle-Zélande
GP Publications Ltd.
Customer Services
33 The Esplanade - P.O. Box 38-900
Petone, Wellington
Tel. (04)685-555   Telefax: (04)685-333

Norway – Norvège
Narvesen Info Center - NIC
Bertrand Narvesens vei 2
P.O. Box 6125 Etterstad
0602 Oslo 6   Tel. (02)57.33.00
Telex: 79668 NIC N   Telefax: (02)68.19.01

Pakistan
Mirza Book Agency
65 Shahrah Quaid-E-Azam
Lahore 3   Tel. 66839
Telex: 44886 UBL PK. Attn: MIRZA BK

Portugal
Livraria Portugal
Rua do Carmo 70-74, Apart. 2681
1117 Lisboa Codex   Tel.: 347.49.82/3/4/5
Telefax: (01) 347.02.64

Singapore/Malaysia – Singapour/Malaisie
See Malaysia/Singapore" – Voir «Malaisie/Singapour»

Spain – Espagne
Mundi-Prensa Libros S.A.
Castelló 37, Apartado 1223
Madrid 28001   Tel. (91) 431.33.99
Telex: 49370 MPLI   Telefax: 575.39.98
Libreria Internacional AEDOS
Consejo de Ciento 391
08009 - Barcelona   Tel. (93) 301-86-15
                Telefax: (93) 317-01-41
Llibreria de la Generalitat
Palau Moja, Rambla dels Estudis, 118
08002 - Barcelona   Telefax: (93) 412.18.54
Tel. (93) 318.80.12 (Subscripcions)
(93) 302.67.23 (Publicacions)

Sri Lanka
Centre for Policy Research
c/o Mercantile Credit Ltd.
55, Janadhipathi Mawatha
Colombo 1   Tel. 438471-9, 440346
Telex: 21138 VAVALEX CE   Telefax: 94.1.448900

Sweden – Suède
Fritzes Fackboksföretaget
Box 16356, Regeringsgatan 12
103 27 Stockholm   Tel. (08)23.89.00
Telex: 12387   Telefax: (08)20.50.21
Subscription Agency/Abonnements:
Wennergren-Williams AB
Nordenflychtsvägen 74, Box 30004
104 25 Stockholm   Tel. (08)13.67.00
Telex: 19937   Telefax: (08)618.62.32

Switzerland – Suisse
OECD Publications and Information Centre
Schedestrasse 7
D-W 5300 Bonn 1 (Germany)   Tel. (49.228)21.60.45
Telefax: (49.228)26.11.04
Librairie Payot
6 rue Grenus
1211 Genève 11   Tel. (022)731.89.50
Telex: 28356
Subscription Agency – Service des Abonnements
Naville S.A.
7, rue Lévrier
1201 Genève   Tél.: (022) 732.24.00
Telefax: (022) 738.48.03
Maditec S.A.
Chemin des Palettes 4
1020 Renens/Lausanne   Tel. (021)635.08.65
Telex: (021)635.07.80
United Nations Bookshop/Librairie des Nations-Unies
Palais des Nations
1211 Genève 10   Tel. (022)734.14.73
Telex: 412962   Telefax: (022)740.09.31

Taiwan – Formose
Good Faith Worldwide Int'l. Co. Ltd.
9th Floor, No. 118, Sec. 2
Chung Hsiao E. Road
Taipei   Tel. 391.7396/391.7397
Telefax: (02) 394.9176

Thailand – Thaïlande
Suksit Siam Co. Ltd.
1715 Rama IV Road, Samyan
Bangkok 5   Tel. 251.1630

Turkey – Turquie
Kültur Yayinlari Is-Türk Ltd. Sti.
Atatürk Bulvari No. 191/Kat. 21
Kavaklidere/Ankara   Tel. 25.07.60
Dolmabahce Cad. No. 29
Besiktas/Istanbul   Tel. 160.71.88
Telex: 43482B

United Kingdom – Royaume-Uni
HMSO
Gen. enquiries   Tel. (071) 873 0011
Postal orders only:
P.O. Box 276, London SW8 5DT
Personal Callers HMSO Bookshop
49 High Holborn, London WC1V 6HB
Telex: 297138   Telefax: 071 873 2000
Branches at: Belfast, Birmingham, Bristol, Edinburgh,
Manchester

United States – États-Unis
OECD Publications and Information Centre
2001 L Street N.W., Suite 700
Washington, D.C. 20036-4910   Tel. (202)785.6323
Telefax: (202)785.0350

Venezuela
Libreria del Este
Avda F. Miranda 52, Aptdo. 60337, Edificio Galipán
Caracas 106   Tel. 951.1705/951.2307/951.1297
Telegram: Libreste Caracas

Yugoslavia – Yougoslavie
Jugoslovenska Knjiga
Knez Mihajlova 2, P.O. Box 36
Beograd   Tel.: (011)621.992
Telex: 12466 jk bgd   Telefax: (011)625.970

Orders and inquiries from countries where Distributors
have not yet been appointed should be sent to: OECD
Publications Service, 2 rue André-Pascal, 75775 Paris
Cedex 16, France.
Les commandes provenant de pays où l'OCDE n'a pas
encore désigné de distributeur devraient être adressées à :
OCDE, Service des Publications, 2, rue André-Pascal,
75775 Paris Cédex 16, France.

75880–7/91

OECD PUBLICATIONS, 2 rue André-Pascal, 75775 PARIS CEDEX 16
PRINTED IN FRANCE
(97 91 08 1) ISBN 92-64-13591-X - No. 45609 1992